T0189385

Advances in Information Security

Volume 103

The purpose of the *Advances in Information Security* book series is to establish the state of the art and set the course for future research in information security. The scope of this series includes not only all aspects of computer, network security, and cryptography, but related areas, such as fault tolerance and software assurance. The series serves as a central source of reference for information security research and developments. The series aims to publish thorough and cohesive overviews on specific topics in Information Security, as well as works that are larger in scope than survey articles and that will contain more detailed background information. The series also provides a single point of coverage of advanced and timely topics and a forum for topics that may not have reached a level of maturity to warrant a comprehensive textbook.

Gautam Srivastava • Uttam Ghosh •
Jerry Chun-Wei Lin

Editors

Security and Risk Analysis for Intelligent Edge Computing

 Springer

Editors
Gautam Srivastava
Department of Mathematics
and Computer Science
Brandon University
Brandon, MB, Canada

Uttam Ghosh
Meharry School of Applied Computer
Sciences
Meharry Medical College
Nashville, TN, USA

Jerry Chun-Wei Lin
Department of Computer Science,
Electrical Engineering and Mathematical
Sciences
Western Norway University of Applied
Sciences
Bergen, Norway

ISSN 1568-2633   ISSN 2512-2193   (electronic)
Advances in Information Security
ISBN 978-3-031-28152-5   ISBN 978-3-031-28150-1   (eBook)
https://doi.org/10.1007/978-3-031-28150-1

This Springer imprint is published by the registered company Springer Nature Switzerland AG
The registered company address is: Gewerbestrasse 11, 6330 Cham, Switzerland

Paper in this product is recyclable.

To Upadhriti and Shriyan. Uttam Ghosh

To Arjun and Krishna. Gautam Srivastava

To my family. Jerry Chun-Wei Lin

Preface

With the proliferation of the development of a broad range of applications and services, cloud computing has captured a lot of attention in the last decade.

Edge computing can play a pivotal role in cloud computing and services by addressing some challenges at the network edge. As an emerging concept, there are still many issues pertaining to security and privacy around data storage, computation, network activity, usage, and location and Intelligent decision-making. There is a need to explore how existing security and privacy measures for cloud computing can be applied to edge computing and how the features like heterogeneity, geo-distribution, and wireless connection can be handled. Moreover, edge servers may have to cooperate in order to complete common tasks, while under different operators. This can exacerbate the problem of security and privacy. Developing our understanding of how to protect user and data privacy in these circumstances is essential to the further development of edge computing.

Security and Risk Analysis for Intelligent Edge Computing collected state-of-the-art findings on security and privacy in edge computing, including studies on architecture and system design, machine learning, secure cryptography, data storage, data processing, and offloading decisions. It aspires to provide a relevant reference for students, researchers, engineers, and professionals working in this area or those interested in grasping its diverse facets and exploring the latest advances in edge computing, with a focus on security and privacy.

In Chapter "A Comprehensive Review on Edge Computing, Applications & Challenges", the authors provide a thorough analysis of edge computing by the authors. The authors describe the practical uses and difficulties of edge computing. Authors of this chapter compare and contrast edge computing with cloud computing in terms of architecture, privacy, and security.

In Chapter "Big Data Analytics and Security Over the Cloud: Characteristics, Analytics, Integration and Security", the authors discuss about what big data analytics is and how beneficial it is to enterprises whose data is constantly increasing. With the emergence of big data, it became natural to move away from traditional silos and towards cloud computing for storage, analysis, fast computation, and accessibility. There is definitely a high synergy between big data and cloud computing, and the cloud offers multiple services to aid the management of big data. However, the cloud does pose some security vulnerabilities which are discussed in this chapter along with the countermeasures that can be taken to prevent those. Lastly, the chapter discusses some open challenges in both big data and the cloud along with the future scope of both which when combined can be absolutely game changing for enterprises.

In Chapter "Federated Learning Enabled Edge Computing Security for Internet of Medical Things: Concepts, Challenges and Open Issues", the authors analyze the services of edge computing and federated learning (FL) in medicine to evaluate the potential of intelligent processing of clinical visual data. The duo of edge and FL allows remote healthcare centers with limited diagnostic skills to securely benefit from multi-modal data. The authors explain the motivation behind integrating FL with edge computing for the Internet of Medical Things (IoMT) and also the areas where the duo is applied for the betterment of IoMT services. Finally, the authors elaborate on the challenges and future directions.

In Chapter "Embedded Edge and Cloud Intelligence", the authors provide a brief comparison of edge and cloud intelligence. The initial sections discuss the advantages and limitations of edge and cloud intelligence. Then, various architectures and methodologies are discussed that can be deployed at several levels to meet the applications and business requirements. Several key performance indicators along with AI model training, optimization, and inference techniques are briefly discussed. Finally, future directions and open challenges are presented.

In Chapter "The Analysis on Impact of Cyber Security Threats on Smart Grids", the authors avoid, mitigate, and tolerate cyberattacks by emphasizing the importance of cyber infrastructure security in conjunction with power application security. Based on the security of both the physical power applications and the supporting cyber infrastructure, a tiered method to risk evaluation is provided by the authors. A classification is offered to emphasize the cyber-physical interdependencies. The controls needed to support the smart grid, and the communications and calculations that need to be protected from cyber-attacks. Next, the authors present current research efforts aimed at improving the security of smart grid applications and infrastructure. Finally, current challenges are identified to facilitate future research efforts.

In Chapter "Intelligent Intrusion Detection Algorithm Based on Multi-Attack for Edge-Assisted Internet of Things", the authors introduce a multi-attack IDS for edge-assisted IoT that combines the backpropagation (BP) neural network with the radial basis function (RBF) neural network. Specifically, they employ a BP neural network to spot outliers and zero down on the most important characteristics of each attack methodology. The findings demonstrate great accuracy in the given multi-attack scenario.

In Chapter "Secure Data Analysis and Data Privacy", the authors look at the current trends in secure data systems in terms of performance and stability. Data loss, theft, or contamination might have severe consequences on the entire firm's activities. The authors discuss the leading data analysis strategies, types, methods, and tools currently utilized in data analysis and how they fit with contemporary computing technologies. They also discuss effective data security types, strategies, and methods, showing situations in which each can be utilized. Also, the authors project the relationship between data security processes and contemporary issues within the organizations and on advancing global patterns. The patterns include advancements in AI, cloud technology, and quantum adoption. We include a discussion on data privacy and how it compares with data security. The content information aims to inform the essence of data protection and guide on ways to deal with oncoming threats to the data.

In Chapter "A Novel Trust Evaluation and Reputation Data Management Based Security System Model for Mobile Edge Computing Network", the authors introduce a multi-attack IDS for edge-assisted IoT that combines the backpropagation (BP) neural network with the radial basis function (RBF) neural network. Specifically, they employ a BP neural network to spot outliers and zero down on the most important characteristics of each attack methodology. The findings demonstrate great accuracy in the given multi-attack scenario.

In Chapter "Network Security System in Mobile Edge Computing-to-IoMT Networks Using Distributed Approach", the authors examine the network attacks associated with vulnerabilities in the Internet of Multimedia Things (IoMT). They also emphasize that the rapid increase in the number of attacks on the IoMT network is due to their inability to run a sophisticated network attack detection system. As the IoMT is resource-constrained (limited CPU, memory, and energy), the authors, however, propose a distributed network attack detection system utilizing multi-access mobile edge computing to protect the IoMT.

In Chapter "Wireless and Mobile Security in Edge Computing", the authors provide an overview of the security concerns and potential security breaches in wireless and mobile edge computing. Here, the authors discuss how such breaches happen and provide recommendations for preventing them. Along with that, the methods for strengthening privacy and security in mobile edge computing are also discussed in this chapter.

In Chapter "An Intelligent Facial Expression Recognizer Using Modified ResNet-110 Using Edge Computing", the authors propose an intelligent computing method based on ResNet-110. The authors implemented this model into experiments and achieved a 96.29% accuracy. The experimental results imply that the proposed methodology outperforms main current recognition models. In order to express our thoughts clearly, the method can be applied to the edge computing environments.

In Chapter "Blockchain Based Simulated Virtual Machine Placement Hybrid Approach for Decentralized Cloud and Edge Computing Environments", the authors explain about use of blockchain technology for virtual machine placement so that identified servers can maintain their utilization limits below the upper threshold which in turn ensures tamper-proof communication among peer servers during placement of virtual machines.

More specifically, the book contains 12 chapters classified into two pivotal sections: The first part presents survey chapters in the areas of Edge Computing, Big Data, Federated Learning, Cloud Intelligence. The second part describes application areas including Smart Grids, Intrusion Detection, Data Analysis, Mobile Edge Computing Networks, the Internet of Medical Things, Wireless Edge Networks, Image Recognition, and Blockchain.

We want to take this opportunity and express our thanks to the contributors to this volume and the reviewers for their great efforts by reviewing and providing interesting feedback to the authors of the chapters.

The editors would like to thank Sushil Jajodia, Pierangela Samarati, Javier Lopez, Jaideep Vaidya (Series Editor-in-Chiefs), Ms. Shina Harshaverdhan, Kate Lazaro, and Rahul Sharma (Springer Project Coordinators), and Susan Lagerstrom-Fife (Springer Senior Publishing Editor), for the editorial assistance and support to produce this important scientific work. Without this collective effort, this book would not have been possible to be completed.

Brandon, MB, Canada   Gautam Srivastava
Nashville, TN, USA   Uttam Ghosh
Bergen, Norway   Jerry Chun-Wei Lin

Contents

About the Editors

Gautam Srivastava Dr. Gautam Srivastava (Senior Member, IEEE) has extensive Editorial Experience including *IEEE Transactions on Fuzzy Systems, IEEE Transactions on Industrial Informatics, Computer Standards and Interfaces, Applied Stochastic Modeling and Business, Information Sciences, IEEE Transactions on Computational Social Systems, ISA Transactions, IEEE Intelligent System,* and *IEEE Sensors.* Dr. Gautam Srivastava received his BSc from Briar Cliff University in Sioux City, Iowa, USA, in 2004, followed by an MSc and PhD from the University of Victoria in Victoria, British Columbia, Canada, in 2006 and 2012, respectively. He then worked for 3 years at the University of Victoria in the Department of Computer Science (Faculty of Engineering), where he was regarded as one of the top Undergraduate professors in Computer Science Course Instruction at the University. From there in 2014, he started a tenure-track position at Brandon University in Brandon, Manitoba, Canada, where he currently is Full Professor. Dr. G (as he is popularly known) is active in research in the fields of Data Mining and Big Data. During his 10-year academic career, he has published a total of 400 papers in high-impact conferences and journals. He has also given guest lectures at many Taiwan universities in Big Data. He currently has active research projects with other academics in Norway, Taiwan, Singapore, Canada, and USA.

Uttam Ghosh joined Meharry Medical College as an Associate Professor of Cybersecurity in the School of Applied Computational Sciences in January 2022. Earlier, he worked as an Assistant Professor of the Practice in the Department of Computer Science at Vanderbilt University, where he was awarded the 2018–2019 Junior Faculty Teaching Fellow (JFTF). Dr. Ghosh obtained his MS and PhD in Electronics and Electrical Communication Engineering from the Indian Institute of Technology (IIT) Kharagpur, India, in 2009 and 2013, respectively. He has post-doctoral experiences at the University of Illinois in Urbana-Champaign, Fordham University, and Tennessee State University. Recently, he has received research funding from the US National Science Foundation (NSF) on edge-cloud interplay for intelligent healthcare. Dr. Ghosh has published more than 100 papers at reputed international journals including IEEE Transactions, Elsevier, Springer,

IET, and Wiley, and in top international conferences sponsored by IEEE, ACM, and Springer. He has coedited and published six books: *Internet of Things and Secure Smart Environments*, *Machine Intelligence and Data Analytics for Sustainable Future Smart Cities*, *Intelligent Internet of Things for Healthcare and Industry*, *Efficient Data Handling for Massive Internet of Medical Things*, *Deep Learning for Internet of Things Infrastructure*, and *How COVID-19 Is Accelerating the Digital Revolution: Challenges and Opportunities*. He is a Senior Member of the IEEE and a Member of ACM and Sigma-Xi.

Jerry Chun-Wei Lin is currently working as the full Professor at the Department of Computer Science, Electrical Engineering and Mathematical Sciences, Western Norway University of Applied Sciences, Bergen, Norway. He has published 500+ research papers in refereed journals with 80+ IEEE/ACM journals and international conferences. His research interests include data mining and analytics, soft computing, deep learning/machine learning, optimization, IoT applications, and privacy-preserving and security technologies. He is the Editor-in-Chief of *Data Science and Pattern Recognition* (DSPR) journal, Associate Editor/Editor for 12 SCI journals including *IEEE Transactions on Neural Networks and Learning Systems*, *IEEE Transactions on Cybernetics*, *IEEE Transactions on Dependable and Secure Computing*, *Information Sciences*, among others. He has served as the Guest Editor for 50+ SCI journals. He is the leader of the well-known SPMF project, which provides more than 190 data mining algorithms and has been widely cited in many different applications. He has been awarded as the Most Cited Chinese Researcher in 2018, 2019, 2020, and 2021 by Elsevier/Scopus and Top-2% Scientist in 2019, 2020, and 2021, respectively, by Stanford University report. He is the Fellow of IET (FIET), ACM Distinguished Scientist, and IEEE Senior Member.

A Comprehensive Review on Edge Computing, Applications & Challenges

Rahul Patel ⓘ, Lalji Prasad, Ritu Tandon, and
Narendra Pal Singh Rathore ⓘ

1 Introduction

Edge computing is revolutionizing the collection, processing, and dissemination of data from billions of devices worldwide. Systems at the network's periphery remain influential due to the IoT's explosive growth and the emergence of new applications that necessitate access to high-performance computing resources in real time. Computing systems are able to speed up the development or support of real applications like video processing and analytics, self-driving cars, artificial intelligence, and robotics thanks to faster networking technologies like 5G wireless. The advent of real-time applications that necessitate processing at the edge will drive the technology forward, whereas early computing goals were to address the costs of bandwidth for data travelling long distances due to the development of IoT-generated data. Seventy percent of IoT data will be processed at the network's periphery by 2025, says the International Data Corporation (IDC) [1]. By 2025, IDC estimates, there will be more than 150 billion connected devices worldwide. The term "edge computing" refers to a type of distributed computing in which data is processed close to the point of creation or consumption, or the "edge." Simply put, Edge computing eliminates the need to send data to a distant data centre and instead stores and processes it near the devices that are collecting it [2]. This is done to ensure that latency issues do not impact the performance of an application and the data, especially real-time data. Because of the explosion in the number of Internet of Things (IoT) gadgets that connect to the internet

R. Patel (✉) · L. Prasad · R. Tandon
SAGE University, Indore, MP, India
e-mail: ritu.tondon@sageuniversity.in

N. P. S. Rathore
Acropolis Institute of Technology and Research, Indore, MP, India

© The Author(s), under exclusive license to Springer Nature Switzerland AG 2023
G. Srivastava et al. (eds.), *Security and Risk Analysis for Intelligent Edge Computing*, Advances in Information Security 103,
https://doi.org/10.1007/978-3-031-28150-1_1

for either receiving information from the cloud or delivering data back to the cloud [3], edge computing has emerged as a way for businesses to save money by performing the procedure locally and thus reducing the amount of data that needs to be processed in a centralized or cloud-based location. Moreover, the operations of many IoT devices produce vast amounts of data. There are many examples of devices that produce data and transmit it across the network, such as those that monitor production equipment on the factory floor or an internet-connected video camera that sends live footage from a remote office. When more and more devices are transmitting data simultaneously, complications arise. As the next big thing in tech, each computing is predicted to upend current computing paradigms. Due to the rapidly growing number of connected devices, ubiquitous connectivity is an urgent requirement. IoT solutions are being widely discussed, re-researched, and adopted because of the widespread availability of several key technologies. Processing the data from devices and performing meaningful interpretation to derive insights and take decisions is the primary challenge in IoT.

With the advent of cloud computing, IoT has been able to gain traction and spread rapidly. It has been crucial to the success of IoT deployments and applications, allowing networks to fulfil their primary storage, communication, and computation needs. In addition, it provides services that can be modified to meet the specific needs of each individual client and use case. Cloud computing's many advantages—including scalability, accessibility, low administrative overhead, payment on demand, and ease of use—have spurred the development of a rapidly expanding industry around the world, one that has already exported billions of dollars.

Increased network latency and jitter, inability to access local contextual information, and lack of support for mobile users are a few areas where cloud computing falls short of meeting the application requirements. There are significant flaws in the cloud computing model with regards to load, real-time, transmission bandwidth, energy consumption, and data security and privacy protection. Edge Computing, Fog Computing, Mobile Computing, Mobile Cloud Computing, and Mobile Edge Computing are all examples of the new computing paradigm that have emerged as a result of these factors. By making use of the hardware, software, and network connectivity at the network's periphery, "Edge Computing" brings essential functions like communication, storage, and computation closer to the devices and users who need them. The primary goals of these computing technologies are to improve upon the limitations of cloud computing by bringing the benefits of cloud computing closer to the user on the network's edge. Figure 1 shows the structure of this article.

With the proliferation of IoT, there is a corresponding increase in network data; edge computing enables devices to perform computations at the network's periphery, on both downstream data on behalf of cloud services and upstream data on behalf of IoT services [4]. Edge computing in this context means processing data using network nodes located between data sources and cloud storage facilities. A smartphone, for instance, acts as a buffer between its user and remote servers. Between a mobile device and the cloud, an edge can be a microdata center or a

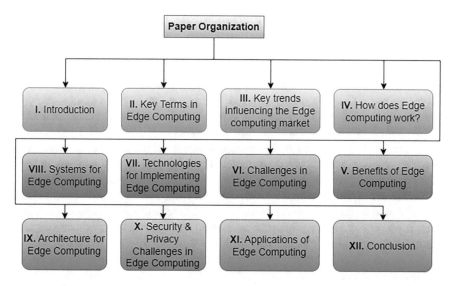

Fig. 1 Paper organization

cloudlet [5], while a gateway connects a smart home to the cloud [6]. With edge computing, tasks like computation, offloading data storage, caching and processing, request distributions, and distribution accommodation can all be handled locally, near the data source.

Computing is moved from the cloud to local places like a user's computer, an IoT device, or an edge server to reduce latency and bandwidth use [7]. In layman's terms, edge computing means running fewer processors in the cloud and moving those processes to local places. By moving computation closer to the network's edge, we can reduce the amount of data that must travel between the client and the server over long distances [8].

Perhaps the most notable development is the increasing prevalence of edge availability and edge services as edge computing adopts new technologies and practices to improve its capabilities and performance. In contrast to the situational nature of edge computing today, the widespread availability of these technologies is expected by 2028, which will change the way the Internet is used and open up new possibilities for edge computing. The proliferation of edge-computing-optimized appliances in the areas of computing, storage, and networking is evidence of this trend. In order to improve product interoperability and flexibility at the edge, more multi-vendor partnerships are needed. One such collaboration is between Amazon Web Services (AWS) and Verizon (VZW) to improve edge connectivity. In the coming years, wireless communication technologies like 5G and Wi-Fi 6 will also influence edge deployments and utilization by enabling virtualization and automation capabilities that have not yet been fully explored, such as improved vehicle autonomy and workload migrations to the edge.

In response to the explosion of connected devices and the resulting deluge of data, interest in edge computing has grown. Despite the immaturity of IoT technologies, the evolution of IoT devices will have an effect on the progress of edge computing in the years to come [2]. One potential future option is the creation of MMDCs, or micro modular data centres. The MMDC is essentially a miniature data centre on wheels, as it contains an entire traditional data centre within a compact mobile system that can be spread out across a city or region to bring computing closer to the location where data is being generated.

The edge of the network is the point at which a device or a local network that contains a device connects to the Internet. The edge is a relative concept, at best. The edge of the network can be anything from a user's computer to the processor in an Internet of Things camera. Users' routers, ISPs, local networks, etc., are also considered to be on the edge [7]. What's crucial to remember is that unlike original servers and cloud servers, which can be located anywhere in the world, the edge of the network is physically close to the device.

When compared to the traditional models of computing—which relied on central servers and peripheral terminals—edge computing takes advantage of the decentralised nature of modern computer architecture. There was a time when individual computers were the norm; in this model, data and programmes resided locally, either on the user's device or in a private data centre.

A more recent innovation, cloud services are a vendor-managed cloud or collection of data centres that are accessible via the Internet and have several advantages over on-premises computing. However, due to the distance between end users and the data centres where cloud services are hosted, cloud computing can introduce latency. To mitigate this, edge computing, which is most computing closer to end users, is becoming increasingly popular. To keep the data from having to travel too far while keeping cloud computing's distributed nature. See how the Edge Computing model stacks up against the Cloud Computing model in Fig. 1.

Consider a building that uses dozens of high-definition Internet of Things security cameras to monitor the premises as an illustration of edge computing in action. It's just cameras sending video data to a remote server in real time. To filter out inactive footage, only those showing motion are stored in the cloud server's database, where they can be viewed at a later time. Due to the high volume of video being transferred, the building's Internet connection is under constant and substantial stress [8]. In addition, the cloud server is under extreme stress from handling the combined amounts of video data from all the cameras. Suppose, instead, that the computation for the motion sensors is offloaded to the network's periphery, to the cameras themselves, with the idea being that they would each use their own processor to run the motion-detecting app, with the output being uploaded to a central cloud server as needed. Since much of the camera footage wouldn't have to make its way to the cloud server, bandwidth usage would be drastically reduced. And with the cloud server's newfound focus on archiving only the most crucial footage, it would be able to efficiently communicate with a large number of cameras. Figure 2 shows a comparison between the edge computing model and the cloud computing model.

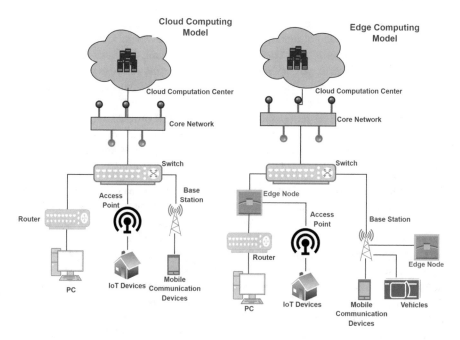

Fig. 2 Comparison between edge computing and cloud computing model

1.1 Edge Computing Versus Cloud Computing

More and more data are being created at the network's periphery. As a result, it makes sense to perform data processing at the network's periphery. As an alternative to the cloud and other computing services, edge computing has several advantages.

Required Response Time Due to the superior computing power of the cloud, which surpasses that of devices at the network's periphery, processing data entirely in the cloud has been shown to be an effective method. With the increasing volume of data being produced at the periphery, the speed at which data can be transported is becoming the bottleneck for the cloud-based computing paradigm [9]. An autonomous vehicle, as another example, will produce 1 Gigabyte of data every second and will need Real-Time processing to be able to make decisions. The response time could not be acceptable if all data had to be sent to the cloud for processing. In addition, the capability of the existing bandwidth and reliability to support such a high concentration of vehicles would be called into question [4]. For faster response times, more efficient processing, and reduced network strain, it is necessary to process the data at the edge in this case.

Needs of IoT Air quality sensors, LED bars, streetlights, and even an Internet-connected microwave oven are just a few examples of the kinds of electrical devices that will soon be a part of IoT and will act as data producers and consumers. In a few years, the number of devices at the network's periphery is expected to grow into the billions. As a result, the volume of raw data generated by them will be so large that it will strain the capabilities of conventional cloud computing. Because of this, the vast majority of IoT data will never make its way to the cloud. In its place, it is used at the network's periphery. The edge data quantity is too high for IoT, which will cause excessive consumption of network and computer resources. Second, cloud computing in IoT will be hampered by the necessity of protecting users' personal information. Last but not least, most IoT end nodes have limited energy resources, and the wireless communication module is notoriously power-hungry. Therefore, it may be more efficient to offload some computing tasks to the edge.

Consuming and Producing Data   Edge devices, in the cloud computing paradigm, typically function as data consumers, such as when viewing a YouTube video on a mobile device. But these days, people are also creating data with their smartphones. An increase in edge-based functionality is needed to facilitate the transition from data consumer to data producer and consumer. It is now common practise for people to record themselves on camera and upload the results to a cloud service like YouTube, Facebook, Twitter, or Instagram to share with friends and family. But the photo or video clip might be quite big, and it would take up a lot of space when uploaded. When this is the case, the video clips should be demised and resized to the right specifications at the edge before being sent to the cloud. Wearable medical technology is yet another illustration. Considering that the information collected by devices at the network's periphery is typically private, processing it at the edge could better protect user privacy than sending raw data to the cloud.

1.2   Key Terms in Edge Computing

Edge Compute   When it comes to supporting various applications, "Edge Compute" is a physical compute infrastructure that sits somewhere between the device and the hyper-scale cloud. Through the use of edge computing, data processing can take place in proximity to the end-user device or data source, thereby avoiding the need to be sent to a remote cloud data centre.

Telco Edge Computing   Customers can run low-latency applications and cache or process data close to their data source to reduce backhaul traffic volumes and costs with Telco Edge Computing, which is distributed computing managed by the telco operator and may extend beyond the network edge and onto the customer edge.

Edge processing that takes place locally. The customer-premises computing resources administered by a network operator for use in running applications and services. These tasks are carried out in a virtualized environment as cloud-based operations across a decentralized edge architecture. Although it takes advantage of the elasticity provided by the edge cloud, sensitive data is kept safely on-premises with on-premises edge computing.

Edge Cloud Edge computing built on top of a virtualized infrastructure and business model. With its ability to handle unexpected increases in end-user activity and its scalability, the edge cloud combines the best features of both cloud and on-premises servers.

Private Cloud Private clouds provide the advantages of public clouds, like scalability and agility, but in a deployment model where computing services are offered over a private network to a set of dedicated users. However, a private cloud's internal hosting of the cloud infrastructure provides greater security and data privacy.

Network Edge This is where enterprise-owned networks such as Wireless LAN or data centers connect to a third-party network like the Internet.

Edge Cloud Edge computing and edge cloud are often used interchangeably, but they actually mean slightly different things. To a large extent, the term "edge computing" is used to describe the actual compute infrastructure that sits somewhere between the end-device user's and a massively scalable cloud, and which in turn supports various applications. Edge Cloud, on the other hand, is a virtualized infrastructure and business model that sits atop the cloud and shares the cloud's scalability and adaptability. It can handle certain spikes and workloads from unanticipated increases in end-user activity, unlike static on-premises servers. It also facilitates scalability during the development and release of new software. Excellent business option, indeed. Once again, savings can be made through increased productivity and scalability.

Where is the Edge However, telco computing is a subset of the edge, which encompasses the entire range of infrastructure between the end device and the cloud or internet. Computing at the network's edge can take place in a variety of places, both within and outside of the main internet infrastructure. Locations in the axis and core networks that serve as aggregation points for smaller networks are also included here [10]. There are three considerations when deciding where to deploy telco's edge computing infrastructure. The telcos current network architecture.

- The virtualization roadmap, which is where you plan data center facilities for network applications.
- Demand and the use cases telco have to cater to.

For instance, cell towers near the customer edge cover a larger area than the street cabinet; however, these were not suitable for the low latency communication required by autonomous vehicles, which require the ability to react instantly to changes in the immediate vicinity and further down the road [11].

Near Edge Components It is possible for WAN infrastructures, such as the hardware colocation at cell towers and cellular switching stations, to coexist with near edge systems, which are part of the infrastructure between the far edge and the cloud layers. Services requiring a high level of computational complexity, such as software-defined wide-area networks (SDWAN), could be hosted at this level.

Far Edge Components This layer is furthest from the overall cloud layer, but it still maintains a relationship with the cloud and its parent near edge components, and consists of processing devices capable of communicating, managing, and exchanging data with cloud and/or near edge appliances. This level serves as the gateway to the end-users or sensor systems. Low latency and support for the vast majority of PAN networks is necessary at this level, as are real-time or safety-critical designs.

2   Key Trends Influencing the Edge Computing Market

Several key factors are driving the growth of Edge computing.

Growth of IoT Rapid growth in the Internet of Things (IoT) adoption is leading to an explosion in data, necessitating the development of centralized cloud computing and storage solutions. Latency, bandwidth, and security issues make cloud computing a questionable option for many businesses, so they're looking instead to edge computing solutions. In order to collect and analyze data from IoT devices in real time, edge computing places the compute closer to the device itself. With a shorter round-trip time to and from the data centre, data is safer and network latency is improved. Edge computing has the potential to enhance. IoT applications, especially those that need instantaneous responses.

Strategic Partnerships and Alliances Network service providers, cloud and data centre operators, traditional enterprise IT and industrial application and system developers, and so on are just some of the newer additions to the still-developing edge computing ecosystem. With a solid foundation laid in the cloud, hyper scalars are expanding their reach to the edge with new solutions like Azure Stack and AWS outposts, as well as their Internet of Things (IoT) offerings. A number of new companies, such as MobiledgeX, Mutable, and Swim.AI, have emerged to fill the void in the market for Paas to manage edge applications. The main competitive strategies used by vendors in this market to gain market share and expertise have been acquisitions and partnerships. Vapor IO, Federated Wireless, Leonardo, MobiledgeX, Packet, and StackPath are just some of the companies that have formed partnerships to promote the spread of edge computing. In order to provide Internet of Things (IoT) edge computing solutions to the manufacturing industry, Litmus Automation teamed up with Siemens. Verizon has teamed up with AWS to provide edge computing on their 5G network by means of AWS's wavelength.

Demand in Key Markets Almost half of the estimated global revenue market share for Edge computing is currently held by the North American region. U.S. manufacturers are investing in connected factories, the most advanced of which may employ Edge computing due to the widespread adoption of IoT, especially in the manufacturing sector. As a result of the widespread adoption of IoT-supported applications and 5G technology, however, Asia-Pacific is expected to experience the fastest growth rates. It is expected that the introduction of 5G connectivity and other investments by the telecom industry in China will be beneficial to the growth of Edge computing in the country. Under its Digital India programme, the Indian government has promised significant funding toward the creation of one hundred "smart cities," which will be run by Internet of Things infrastructure. Edge computing consortiums have been formed in Japan to aid in the country's efforts to advance its computing products and converge on a common set of industry standards.

2.1   Working of Edge Computing

Data in conventional enterprise computing is generated at the client endpoint, such as a user's computer, whereas in edge computing, this is not the case. Using the company's local area network (LAN), the data is transmitted to a wide area network (WAN) like the Internet, where it is stored and processed by an enterprise application. The completed tasks' results are then sent back to the user interface. For the vast majority of standard enterprise software, this tried-and-true client-server architecture continues to deliver the goods.

The issue is that the growth rate of Internet-connected devices and the volume of data produced by those devices and used by businesses is much faster than that of the infrastructures supporting data centers. According to Gartner, by 2025, non-DC data will account for 75% of all enterprise data. Since the global Internet is already prone to congestion and disruption, the prospect of moving so much data in situations that can be time- or disruption-sensitive places enormous strain on it.

As a result, IT architects have begun to relocate data storage and processing closer to the physical edge of their networks, or where the data is actually being created. If you can't move the data to where the data centre is, move the data centre to where the data is. The idea of Edge computing has its origins in remote computing concepts that date back decades, such as home offices and satellite locations. In some cases, decentralizing computing resources was preferable to relying on a centralized server [2, 12]. In order to process data locally, edge computing places storage and servers close to the source of the data, requiring only a partial rack of equipment to run on the remote LAN. Computing equipment is often deployed in shielded or hardened enclosures to prevent damage from dust, moisture, and temperature swings. Normalizing and analyzing the data stream for business intelligence is a common part of the processing phase, and only the results of the analysis are relayed to the main data centre. Business intelligence, as a concept, is extremely malleable. In a retail setting, for instance, surveillance footage from the sales floor could be combined with hard data on what actually sells to infer the optimal product configuration or consumer demand. Predictive analysis is another example that can help with equipment upkeep and repair before any problems

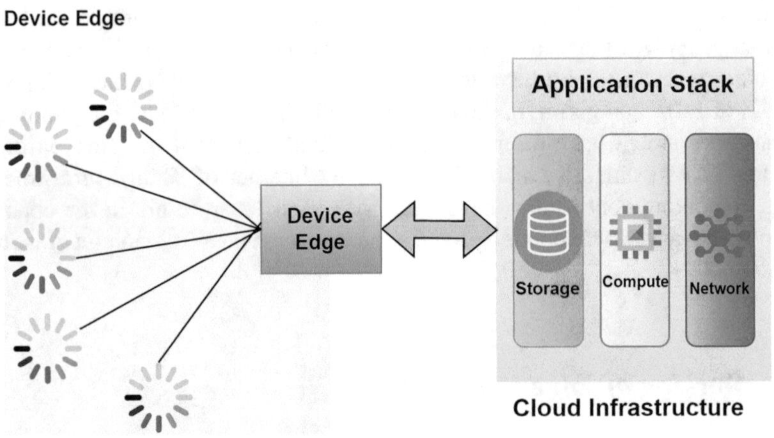

Device Edge

Fig. 3 Device edge

have even arisen. Still other instances are typically paired with utilities like water treatment or electricity generation to guarantee the efficiency of machinery and the reliability of output.

2.2 Forms of Edge Computing

Device Edge Figure 3 is a diagram of a device on the edge, illustrating how this model moves computing closer to the end user by leveraging preexisting infrastructures like Amazon Web Services' green grass and Microsoft Azure's Internet of Things.

Cloud Edge As can be seen in Fig. 4, this Edge computing model is essentially an expansion of the public cloud. The use of caches at geographically dispersed edge locations to speed up the delivery of static content is a paradigmatic example of this approach in content delivery networks. For example, Vapor IO is quickly becoming a major player in this field. They're making an effort to set up cloud-edge infrastructure. VaporIO offers several products, including the self-monitoring Vapor Chamber. They're equipped with sensors that allow vapour software or VEC, short for vapour edge controller, to evaluate and assess their performance in real time.

The primary distinction between the device edge and the cloud edge is the deployment and pricing models that are tailored to each respective use case. It's possible that utilizing both models could be useful in certain situations.

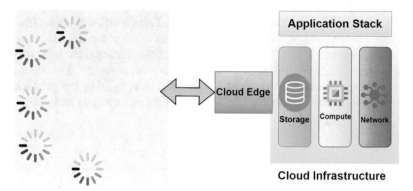

Fig. 4 Cloud edge

3 Benefits of Edge Computing

Here are some of the key benefits of Edge computing.

3.1 Reduced Latency

It becomes less useful information as processing time increases. Time is of the essence in the context of the autonomous vehicle, and the vast majority of the information it gathers and needs is rendered useless after a few seconds at most. In a digital factory, where data consistency is paramount, every millisecond counts thanks to intelligence-based systems that constantly track every step of production. There isn't always enough time for data to travel both ways between clouds. Critical incidents and equipment breakdowns necessitate immediate data analysis. By keeping data processing close to the source, where new data is being generated, we can eliminate latency and speed up responses. In doing so, your data becomes more pertinent, useful, and actionable. Your business's overall application and service performance can benefit from edge computing because it lowers network traffic loads.

3.2 Improved Security

Critical business and operating processes that rely on actionable data are extremely vulnerable when all of that data must eventually see to its cloud analyzer through a single pipe; as a result, a single attack can disrupt an entire company's operations.

Distributing data analysis tools throughout an organisation also means spreading the associated risk around. Edge computing may increase the attack surface for hackers, but it also mitigates their impact on the business. From a different angle, less data transferred means less data to be intercepted. Due to the increased use of mobile computing, businesses are at a much higher risk of being breached because company devices are frequently taken outside the secure firewall perimeter. The security of an on-premises business is maintained even while data is being analysed locally. Edge computing also aids businesses in resolving data sovereignty and compliance concerns at the regional level.

3.3 Reduced Costs

Given that not all data is created equal and not all data has the same value, it seems unreasonable to allocate the same resources toward transporting, managing, and securing all data. Some information is indispensable to running your business, but other pieces of data could be tossed aside. By keeping as much data as possible in your edge locations, you can save money on bandwidth costs that would otherwise be required to connect all of your locations and improve your data management. Instead of trying to do away with the cloud entirely, Edge Computing aims to supplement it. It's all about streamlining your data flows to cut costs wherever possible. Data created at the edge must be stored there at least temporarily before it can be sent to the cloud, where it must be stored again, which is why Edge Computing is useful for reducing re-redundancy. Reducing redundant storage also decreases redundant expenditures.

3.4 Improved Reliability

The Internet of Things (IoT) encompasses a vast global landscape, some of which consists of rural, low-connectivity areas. Increased dependability [13] is achieved when edge devices are able to locally store and process resulting data. These days, micro data centres can be purchased ready-to-use in a prefabricated form and used in virtually any setting. This means that smart devices can continue to function normally despite losing connectivity to the cloud temporarily. Furthermore, there is a cap on the total amount of data that can be transferred at once, which varies from one site to the next. Even if your bandwidth needs haven't been fully realized just yet, the exponential growth in generated data will eventually strain the capacity of many businesses' bandwidth infrastructure [11].

3.5 Scalability

Despite the theory's apparent contradiction, the idea that edge computing provides a scalability advantage makes sense. In most cases, data must be transmitted to a centralized data centre before it can be used in a cloud computing architecture. Adding on to or even making minor adjustments to specialized data centres is a costly endeavor. And rather than having to wait for the coordination of efforts from staff located in different locations, IoT devices can be implanted together with their processing and data management tools at the edge in a single implantation.

4 Challenges in Edge Computing

4.1 Programmability

In cloud computing, users write their own programs and then upload them to the cloud, with the cloud service provider in charge of determining whether or not the actual processing takes place in the cloud. They don't understand how the programme works, or they only understand it in part. One advantage of cloud computing is that the underlying infrastructure is hidden from the end user. Edge nodes, on the other hand, are likely to be heterogeneous platforms, and in Edge computing, computation is offloaded from the cloud. As a result, the programmer faces substantial challenges when attempting to create an application that can be deployed in the Edge computing paradigm, as the runtime of these nodes varies widely.

4.2 Naming

One key premise of edge computing is that there are an enormous number of things; however, at the edge node, many applications are simultaneously active, and each application, like any computer system, has a structure regarding how the service is provided. Edge computing's naming scheme is important for many reasons, including but not limited to: programming, addressing, thing identification, and data communication. The Edge computing paradigm has not yet had a standardized naming mechanism built for it, though. Professionals working at the edge often need to become fluent in a variety of network and communication protocols so that they can exchange data with the various components of their system. Edge computing requires a naming scheme that can accommodate the mobility of things, highly dynamic network topology, privacy and security protection, and scalability aimed at an extremely large number of unreliable things. When applied to the ever-changing Edge network, the tried-and-true methods of naming fail.

4.3 Data Abstraction

The edge operating system can host a wide variety of data consumers and service providers that exchange information via the service management layer's application programming interface (API). In the paradigms of wireless sensor networks and cloud computing, data abstraction has been the subject of extensive discussion and study. However, this problem becomes more difficult in Edge computing.

In a smart home, for instance, almost all of the things that are on report data to the edge OS, and that's before we even consider the large number of things that would be deployed throughout the home as a result of the Internet of Things. However, most devices at the network's periphery only report sensed data to the gateway on a periodic basis. If the thermometer were to report the temperature every minute, the actual user would probably only look at it several times a day.

4.4 Service Management

Differentiation, extensibility, and isolation are four fundamental features that must be supported for service management at the network's edge to ensure a dependable system.

Differentiation Multiple services, including those for the smart home, will be implemented at the network's periphery as IoT deployment increases rapidly. Different services will be given higher priority at different times; for instance, services dealing with the diagnosis and repair of malfunctioning components will be processed more quickly than those dealing with more frivolous pursuits like providing amusement.

Extensibility In contrast to a mobile system, in which new devices can be added to an existing service with little effort on the part of the owner, IoT devices may present a significant challenge to network edge extensibility. As an alternative, can the existing service readily adopt a new node when something needs to be replaced due to wear and tear? An adaptable and scalable service management layer in the Edge OS is the key to fixing these issues.

Isolation Another problem at the network's periphery would be its isolation; when a mobile operating system's app fails or crashes, the whole system typically does as well. Locks and token rings are two examples of synchronization mechanisms that could be used to manage a shared resource in a distributed system. However, this could become more of a challenge in a clever Edge OS. For instance, light regulation is just one area where several applications might need access to the same set of data. If a user's lights stopped responding because an app stopped working, the edge operating system shouldn't crash entirely.

4.4.1   Application Distribution

The computing centre is moving from the cloud to edge nodes, making the distribution of individual applications to different Edge nodes a crucial issue that has a direct impact on the efficacy and scalability of Edge computing.

In Edge computing, the task of application distribution involves breaking down applications into smaller pieces that can be run on a variety of Edge nodes while still retaining their original applications' semantic information and computing resources, energy efficiency, and response time. This method of application distribution is intended for isomerism nodes, which are not optimal for use in edge computing.

4.5   Scheduling Strategies

It is anticipated that Edge Computing's scheduling strategies will maximize resource utilization, shorten response times, lessen energy consumption, and speed up the processing of tasks. Edge computing scheduling strategies must coordinate computing tasks and resources between nodes in the same way that traditional distributed systems do. At the same time, it's important to remember that, like cloud computing, your computing resources are likely to be of varying types and qualities. In addition, Edge computing's limited computing resources mean that, in contrast to the more open environment of the cloud, managing those resources effectively is one of the biggest hurdles that users must jump.

Edge computing's scheduling strategies, like those for data, computing, storage, and networks, need to be tailored to individual applications because of the diverse nature of these resources. In addition, the approaches must account for the fact that there may be several distinct kinds of applications [9]. Scheduling strategies should maximize the use of scarce resources on Edge nodes to boost application performance and efficiency.

4.6   Optimization Metrics

To do so with Edge computing. Since each layer has its own set of computational limitations, dividing up the work between them can be difficult. Allocation strategies to finish a workload include completing as much as possible on each layer and dividing the workload evenly across all layers. To pick the best allocation method, it is necessary to analyze several crucial indicators.

Latency When measuring the effectiveness of an interaction, application, or service, latency is a key factor. High computational power is made available by cloud computing's server infrastructure [4]. They're great for tasks that require a lot of processing power but not a lot of time, like image and voice recognition. While processing time is a major factor, long-term delays can have a major impact on the functioning of real-time or interaction-intensive programs.

Bandwidth From the perspective of latency, high bandwidth can shorten the time it takes to send data, particularly when it involves large data such as video for short-distance transmission, and we can set up high bandwidth wireless access to send data to the edge. However, if the workload can be processed locally, there may be significant latency savings compared to doing the same thing in the cloud. Moreover, bandwidth is conserved in the connection between the edge and the cloud [4]. Since the transmission section is so brief, reliability is increased as well. However, since the edge cannot meet the computation demand, data is processed at the edge, resulting in a significantly smaller upload data size.

Energy When considering the endpoint layer, offloading workload to the edge can be treated as an energy-free method because batteries are the most precious resource for things at the edge of the network. Computing, energy, and transmission energy consumption must be balanced. The transmission energy overhead is affected not only by the network signal strength but also by the data size in the available bandwidth; we prefer to use Edge computing only if the transmission overhead is smaller than computing locally.

Cost To put it another way, from the perspective of a service provider like YouTube, Amazon, etc., Edge computing reduces latency, energy consumption, and potential bottlenecks, while potentially increasing throughput and enhancing the user experience. This allows them to make more money while doing the same amount of work. Service providers must pay to create and maintain the elements of each tier. Providers can charge users based on the location of the data, maximizing the use of local data at each layer. In order to ensure the service provider's profit and the acceptance of users, new pricing models will need to be developed

5   Technologies for Implementing Edge Computing

5.1   Cloudlet

The issue of end-to-end responsiveness between a mobile device and the associated cloud may be solved by using cloudlets. A cloudlet is a small-scale cloud data centre with enhanced mobility that sits at the network's periphery. Accessible to nearby mobile devices, it is a trusted resource-rich system or cluster of systems with strong connections to each other and the internet. Cloudlets are the "middle tier" in a three-tier hierarchy architecture designed to improve response times. The cloudlet's purpose is to reduce ping times by making robust computing resources available to mobile devices, thereby enabling the deployment of resource-intensive and interactive mobile applications [14].

In order to perform computations for the cloud, mobile devices connect to a nearby cloudlet. An important distinction between a cloud and a cloudlet must be made here.

- A cloudlet is much more agile in its provisioning because associated edge devices are highly dynamic and mobile.
- A good handoff is needed to support the seamless mobility of edge devices.

A mobile node will first establish a connection to a nearby cloudlet before connecting to the cloud itself. Through the use of cloudlets, the network's data load can be lightened and the delay in responses to mobile nodes can be minimized.

5.2 Micro Data Center

Recent years have seen a surge in demand for low latency access to data processing and storage due to applications like the Internet of Things (IoT), content delivery, and fifth generation mobile networks (5G). Traditional centralized data centres, like the ones used by Amazon Web Services and Microsoft Azure, weren't intended for those kinds of workloads. These massive data centres are essential to the success of cloud computing, as they enable economies of scale and facilitate the movement of computing closer to the edge, but they are not able to move data processing close enough to end users for certain distributed workloads. Micro data centres (MDCs) can be as small as a single 19-inch rack or as large as a 40-foot shipping container, and they contain all the computing, storage, networking, power, and cooling resources necessary to support a specific workload. There are places where larger data centres simply won't work, but micro data centers can be set up just about anywhere. As a result, businesses are able to affordably address the issues of low-latency data processing and storage. With the help of MDC, edge computing can be implemented in the real world. The "what" is edge computing, and the "how" includes micro data centres and the underlying network infrastructure.

5.3 Multi-Access Edge Computing

The capabilities of cloud computing can be expanded by bringing it to the edge of the network, and this is what Multi-Access Edge Computing (MEC) does. The Mobile Edge Computing (MEC) standard was developed as part of an initiative launched by the European Telecommunications Standards Institute (ETSI) to facilitate the deployment of mobile network edge nodes. However, it has evolved to incorporate the fixed, and even converged, network. In contrast to traditional cloud computing, which occurs on remote servers located far from the user and the device, MEC enables processes to occur in base stations, central offices [15], and other aggregation points on the network. Mobile edge computing (MEC) is a technique for improving end-user experiences by offloading cloud computing tasks to users' own local servers, thus reducing network congestion and increasing latency.

It's true that MEC can be used to boost the functionality of preexisting applications like content delivery and caching, but it's also emerging as a crucial enabler for brand new uses. New revenue-generating use cases for 5G will be driven by MEC, and efficiency gains for Telcos in delivering highly distributed high throughput content may also be possible. New types of consumer, business, and medical applications and services are made possible by MEC because of its cutting-edge features like low-latency proximity, high-bandwidth, and real-time insight into radio network information and location awareness [14]. The use of MEC to manage video streaming services in the context of smart cities also appears to be a viable option.

5.4 Fog Computing

As can be seen in Fig. 5, the new technology known as fog computing is very similar to edge computing, but it integrates with the cloud. Fog computing is distinct from edge computing because it offers mechanisms for inter-device and inter-network resource management and security. The edge is where servers, apps, and micro clouds are located in an edge architecture [16]. Unlike edge computing, which is defined by its absence from the cloud, fog computing works in tandem with the cloud. In both fog computing and edge computing, the middle layer is depicted, but in fog computing, the middle layer interacts with the cloud, whereas in edge computing, it is depicted as being closer to the edge device. The open fog consortium is primarily responsible for coordinating the standardization of fog computing; its mission is to persuade standardization bodies to develop specifications that will allow edge IoT systems to communicate safely and smoothly with other edge and cloud services.

5.5 Mobile Cloud Computing

Application delivery to mobile devices is the focus of mobile cloud computing. This remote deployment option makes it possible to rapidly iterate and test these mobile applications. Thanks to cloud services, mobile cloud applications can be developed and updated rapidly. Applications that would not normally be supported can now be made available to users thanks to their portability across a wide range of platforms, computing tasks, and data storage. The end goal of MCC is to facilitate the execution of high-quality mobile applications across a wide variety of mobile devices [17]. Both mobile network operators and cloud service providers can benefit from MCC. The term "mobile cloud computing" (MCC) refers to a type of cloud computing that uses the unified elastic resources of multiple cloud and network technologies

Fig. 5 Fog computing

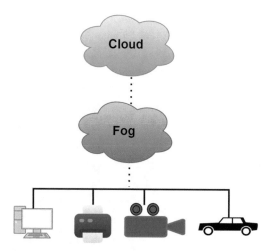

to provide unlimited functionality [18], storage, and mobility to a large number of mobile devices over an Ethernet or Internet connection, regardless of the devices' underlying operating systems or environments, on a "pay as you go" basis.

5.6 Edge Virtual Machines

The storage and compute resources, the power, and the virtual machine are moved to a location closer to the end-user, making up an edge virtual machine. A virtual machine (VM) is a software-defined computer that boots into its own operating system on top of another OS on a host server. The resources of a single physical machine can be shared by numerous virtual machines. All of these VMSs are completely self-contained and capable of performing all the duties of any OS in an IT infrastructure on their own building blocks [19]. With the deployment of computing and rising demand of virtual machines at the edge, the need for Edge Computing to bring computing capability to the very edge of the network has arisen. Edge virtual machines (VMs) and containerized applications hold the promise of a much more rapid and efficient network infrastructure.

5.7 Edge Containers

Edge containers are distributed computing resources that are placed as close to the end-user as possible in order to reduce latency, save bandwidth, and improve the digital experience. By shifting intelligence to the network's periphery, or "edge," organizations can reduce network costs and improve response times [20]. Edge

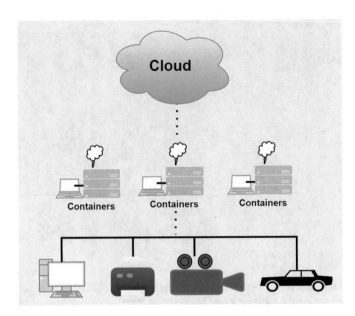

Fig. 6 Edge containers

containers make this possible by allowing organizations to decentralize services by moving key components of their application there.

Containers are well-suited to edge computing solutions due to their ease of deployment, software package distribution, and application containerization. As shown in Fig. 6, unlike conventional cloud containers, edge containers can be deployed in parallel to geographically diverse points of presence, increasing availability. Edge containers are located at the network's periphery, much closer to the end-user, while cloud containers are hosted in far-flung continental or regional data centres.

5.8 Software-Defined Networking (SDN)

With edge computing, the computational infrastructure is moved closer to the data source, and the computing complexity rises proportionally. By providing automatic Edge device reconfiguration and bandwidth allocation, SDN simplifies network complexity while also providing a cost-effective solution for virtualizing the network's periphery. By using software-defined networking, edge devices can be deployed and configured with minimal effort [9]. IoT, smart home, and smart city system security are all areas where SDN shows great promise [21].

5.8.1 Similarities and Differences Between Edge Paradigms

It is the goal of all edge paradigms to bring some of cloud computing's power to the edge. All of these solutions contribute to the multitenant virtualization infrastructure that facilitates location-aware provisioning and quick access to on-demand computing power in close proximity. The most important similarity is that all of these methods can aid mobility in some way or another. They take device mobility into account and either have management mechanisms for it built into the application or employ virtual instances of devices to make this possible [14]. Next, the overall architecture provides the necessary support to make edge paradigms operate just like a miniature cloud. Elements of the network function independently and in a decentralized manner, able to provide services and make decisions without any central authority. They can also work together to rely less on the centralized cloud infrastructure.

There are some fundamental distinctions between the various Edge paradigms in how they achieve the goal. While most MEC deployments are focused on making 5G a reality, fog nodes can offer their services to third-party applications that already have their own infrastructure in place (such as servers, gateways, access points, and so on). The MCC service provider is decided by the main difference between the management and deployment of data centres, as MCC is highly distributed and device instances can perform their own service provisioning. The selection of service providers also affects the types of applications that can be used. Even for fog computing, MEC enables operators to work closely with other third-party service providers, allowing for comprehensive testing and the possibility of customized integration. In contrast, the MCC paradigm offers services that facilitate the use of distributed execution mechanisms, but which are unrelated to virtualization. Very low-power devices will benefit greatly from this [9]. To a large extent, the security and privacy methods developed for one paradigm can be applied to the other.

6 Systems for Edge Computing

We discuss a few of the more popular systems currently available; this is by no means an exhaustive catalogue, but rather an indicative sample.

6.1 Open-Source Systems

Here are some of the prevalent open-source systems in the market today as shown in Fig. 7 are:

Apache Edgent It's a programming model that allows for real-time processing of data streams, right at the edge, in the devices and gateways. Edgent decides whether

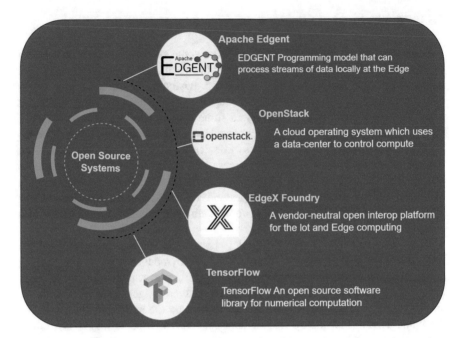

Fig. 7 Open-source systems

or not data should be stored or analyzed in the edge device or in backend systems. With Edgent's help, the app can adapt, sending only the most pertinent information to the server rather than the continuous stream of raw data. This could drastically cut down on the data being sent and stored on the server [9].

OpenStack It's an OS for the cloud, which means it manages computational, storage, and communication assets in a centralized data centre. In addition to a user-friendly web interface, it also includes dashboard management tools. Open Stack's distributed software enables edge computing by providing support for virtual machines and container technologies. Open Stack's basic infrastructure can be deployed at edge devices.

EdgeX Foundry For Internet of Things (IoT) and edge computing, it is a vendor-agnostic, open Interop platform. It is a Linux Foundation-hosted interoperability framework for the entire hardware or operating system stack. Those with an interest in edge computing are free to use existing communication standards and their own proprietary innovations to freely collaborate on Internet of Things solutions. Ajax Foundry prioritizes industrial IoT by catering to the unique requirements of IoT communication protocols while also utilizing cloud computing best practices.

TensorFlow It's an open-source library for performing numerical computations with data flow graphs, and its adaptable design lets you use a single application programming interface (API) to distribute workloads across multiple CPUs or GPUs

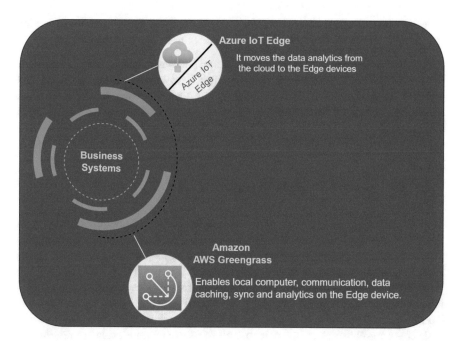

Fig. 8 Business systems

on a desktop server or mobile device [22]. Linear regression, neural network, SVM, K-means, and many more widely used machine learning algorithms are all available in TensorFlow. There is a wide variety of languages supported by the API. Most importantly, it is applicable to a wide variety of computing environments, from desktop to mobile to edge devices, thanks to its support for heterogeneous platforms

6.2   Business Systems

Figure 8 shows that in addition to open-source solutions, cloud service providers like Azur-IoT, Edge from Microsoft, Google Cloud IoT, and Eatables Green Grass from Amazon offer business systems that enable advanced data analytics and artificial intelligence at the edge of the network.

Azure IoT Edge   The Azure IoT Edge consists of three parts: the IoT modules, the IoT as runtime, and the cloud-based interface, and it shifts data analytics from the cloud to the Edge devices. Through the IoT Edge runtime, edge devices can manage communications and operations using cloud logic. Meanwhile, this device supports multiple IoT edge modules, each of which can be run as its own docker-compatible container and used to execute either Azure services, third-party services, or user-supplied code. Edge devices' workloads can be distributed and monitored via the IoT cloud interface.

Amazon AWS Green Grass Edge computing software is what makes it possible for the device to function as a local computer, to communicate, to cache and synchronize data, and to perform analytics. Local execution of AWS lambda, messaging, device shadows, and security are all made possible by the Green Grass Core runtime.

7   Architecture for Edge Computing

In most cases, edge solutions involve a distributed architecture with multiple layers that distributes work between the edge, the cloud, or the network and the enterprise. Some of the terms found in edge computing architectures, such as OT (which stands for "operational technology"), are taken directly from the Internet of Things (IoT) framework. The Internet, personal computers, and Wi-Fi access points are now fully integrated into operational technology (OT) environments. Over the next decade, the transformation and convergence of IT and OT technologies have the potential to deliver enormous value; however, not all of its computing levels are required in every scenario. Devices at the network's periphery can and will keep running even if they lose contact with the cloud or are in an otherwise disconnected environment, like on an oil rig platform. Due to the exponential growth in both the number of connected devices and the volume of data they generate, traditional Internet of Things (IoT) architectures will inevitably experience bottlenecks and delays. This is where the edge comes in. In order to overcome these obstacles, edge computing relocates the process to the network's periphery. Have a look at Fig. 9 for a high-level overview of the edge computing architecture's four layers or regions.

7.1   Edge Device

Due to the capabilities of edge and IoT devices to perform analytics, apply AI rules, and even store some data locally to support operations at the edge, these devices may be able to handle analysis and real-time inference independently of the edge server or enterprise layer. Due to cost and size constraints, edge devices typically have low computing power, but they can still use any SaaS they want.

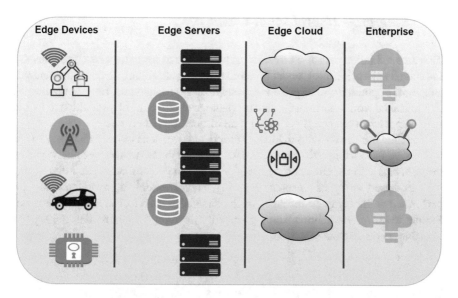

Fig. 9 Architecture of edge computing

7.2 Edge Servers

Application distribution occurs via Edge Servers, which stay in constant contact with the devices via agents. These edge servers monitor the plethora of devices and, if more than inferencing is required, receive data from the devices for processing [23]. These are remote, racked computers with 8 or 16 or more cores of computer capacity, 16 GB of memory, and several hundred GBs of local storage used in places like factories, stores, hotels, and banks.

7.3 Edge Cloud

Edge Cloud the edge cloud, also known as a micro data centre, is the product of recent advancements in networking technology and can be thought of as a local cloud that devices can use to communicate with one another. The term "Edge network" is sometimes used by the telecom industry to describe this type of network. With the advent of 5G, it also helps with bandwidth problems and reduces latency because it shortens the distance data from the devices must travel. These facilities also provide extra space for analytical and data model storage.

7.4   Enterprise Hybrid Multi-Cloud

This area, which is part of a larger hybrid multi-cloud for enterprises, provides the traditional enterprise-level model for storage and management, device management, and, most importantly, enterprise-level analytics and dashboards. Edge computing is distinguished from traditional server and storage architectures by the drive to locate these elements at the periphery rather than in a central data centre.

The time it takes for data to be transferred and decisions to be made are reduced by moving processing tasks from the data centre to the network's periphery, where they are closer to the end user. At the edge, you can perform real-time inferencing, but rule-based analytics, machine learning, and/or model improvement must be performed at the layers to the right, such as the edge server or even the enterprise. Similar to how there is limited storage on edge devices, AI/data models are kept in the edge server or edge cloud.

8   Security and Privacy Challenges in Edge Computing

8.1   Security in Edge Computing

The following are some of the most pressing issues that must be addressed in order to guarantee confidentiality and privacy. The additional security mechanism shouldn't make the system slower or require more storage space than is necessary. Cash management techniques used by edge computing mechanisms are vulnerable to side-channel attacks. It is therefore important to take precautions against the disclosure of confidential information. The increased number of devices means that applications that stream data continuously for monitoring and management purposes typically have a large volume of data to process. Packet filtering mechanisms must be implemented to detect the anomaly, which may call for more storage space and processing power. The large number of users that share the edge computing environment's resources and applications makes it challenging to manage user identity, scalability, monitoring, performance, data security, and considering threats from insiders.

It is entirely up to the needs of the user or the application to determine the balance that must be struck between speed and safety. There must, therefore, be well-defined performance standards that do not compromise safety. Data storage and recovery issues may arise because of the edge platform's potentially high data throughput but relatively low storage capacity, depending on the application and user requirements. Through the use of virtualization services, edge computing enables multitenant application access [14]. This leaves the system vulnerable to insider attacks and data breaches as a result of the increased number of users. Implementing a uniform policy for controlling access to these devices is a time-consuming process, especially when one must also take into account the devices' inherent mobility.

The various components of providing end to end security are as follows:

Network Security Administrator-performed configurations and network management data must be encrypted and kept separate from regular traffic. In light of the dispersed nature of edge devices and the consequently high cost of upkeep [24], this is essential. However, a new networking concept known as software-defined networking (SDN) has come to the rescue, with benefits including, but not limited to, easier management and implementation, greater network scalability, lower maintenance costs, traffic prioritization and isolation, and control over network resources, collaboration, and sharing.

Data Storage Security As with cloud computing, the edge platform stores data off-site, making it difficult to guarantee data integrity due to the high likelihood of data loss and tampering. If an attacker gains access to the data on just one device, they can use it for whatever purpose they like. So, it's important to make room for auditing data storage. Infrastructure providers often offer auditing services from an outside party, and application end users should be aware of these policies. Encryption methods can be utilized to check for untrusted network entities and permit the user to verify the stored data while maintaining its integrity, verifiability, and confidentiality. Since not all data has to be fundamentally present on all available storage resources, low latency and dynamic access are essential.

Data Computation Security All of the data processing that takes place on Edge servers and devices must be safe and auditable. Using data encryption techniques, which prevent data visibility to any hackers or attackers, ensures the security of computations. For this reason, micro data centres can delegate some of their processing to larger data centres. To build trust between the two parties, it is necessary to have a system in place to confirm the accuracy of the calculations [24]. The user should be able to independently verify the computational accuracy. One way to guarantee the safety of data is to encrypt it before sending it from endpoints to edge data centres or between data centres.

End Devises Security Most end devices are less robust than those used in the development process, and they have restricted access to their environments, making them easy to manipulate. Any malicious actor can try to compromise a system and turn it into a rogue node in order to steal sensitive management information from the network. They can disrupt regular operation by altering the device's data and increase data access with the help of fabricated information. They can also use the client machine to spread false information throughout the network, which could lead to inconsistencies and abnormal operation. Therefore, the right amount of care should be taken to guarantee the end device is safe.

Access Control Since the edge platform is distributed and decentralized, a good access control policy accesses a defensive shield to mitigate unauthorized device and service access [14], so enforcing access control mechanisms can provide dual benefits of security and privacy. Interoperability and cooperation between micro data centres from different service providers and in different geographical locations can be achieved through the use of access control. To achieve the desired outcomes of the design, support mobility, low latency, and interoperability, a strong access control mechanism is necessary.

Intrusion Detection Methods of intrusion detection aid in the detection of anomalies in devices and the identification of malicious data entries. They offer tools for performing packet inspection, which aids in the early detection of attacks like denial-of-service, integrity attacks, and data flooding, and can be used to conduct in-depth investigations and analyses of the network's devices' behavior. Implementing intrusion detection for edge platforms is difficult because of the need to meet competing demands for scalability, mobility, and low latency [25].

8.2   Privacy in Edge Computing

Privacy in edge computing can be considered in three ways user privacy, data privacy, and location privacy.

User Privacy A user's login information and how often they access the data are both important pieces of information that must be kept secret. If a hacker can see a user's habits, they may try to pose as a legitimate user to gain access to sensitive data, so protecting their anonymity is essential.

Data Privacy In spite of the existence of data privacy-preserving techniques at the edge and cloud levels, it is a design challenge to make them accessible on the resource-constrained end device. Thus, the security of the network is improved and hackers are discouraged from breaking in.

Location Privacy Since edge data centres can be uniquely dispersed, meeting the need for location privacy is also of paramount importance. Since geolocation information can aid in understanding facts about the natural world. Planning is essential before selecting edge data centres. Maintaining location privacy without increasing computational overhead is of utmost importance, and identity obfuscation is one method for doing so, even when a user is in close proximity to an edge device

9   Applications of Edge Computing

9.1   Transport and Logistics-Autonomous Vehicles

One of the first uses for autonomous vehicles will be autonomous platooning, most likely among truck convoys. In this scenario, traffic is reduced and fuel costs are reduced as a group of trucks travels in a convoy behind one another. Due to the ultra-low latency enabled by edge computing, it will be possible to eliminate the need for drivers in all trucks except the lead truck [26].

9.2 Cloud Gaming

Latency plays a crucial role in the emergence of a new genre of gaming that transmits the action of a game in real time to players' devices. Data centres handle game processing and hosting [27]. In order to provide a seamless and engaging gaming experience with minimal latency, cloud computing providers are aiming to locate their edge servers as close to players as possible.

9.3 Healthcare

Glucose monitors, health tools, and other sensors used in in-hospital patient monitoring are either not connected or generate massive amounts of raw data that must be stored in a remote cloud service. The edge in the hospital could process data locally to maintain data privacy, which raises security concerns for healthcare providers. Additionally, with Edge's real-time alerts, doctors can be made aware of any abnormal trends or behaviors among their patients, and comprehensive dashboards can be built for each individual [28].

9.4 Retail-Contextual Digital-out-of-Home Advertising

Because targeted advertising is more profitable than non-targeted advertising, advertisers are constantly looking for new ways to customize Ad Words. Data on each must be captured and processed in real time for digital out-of-home advertising to deliver a contextualized ad-word. Personal data can be analyzed locally to create timely and relevant advertising by moving this processing to the edge, either on-premises or in the network.

9.5 Retail In-store Immersive Experiences

Stores are increasingly turning to digital media with interactive elements to lure customers in and keep them from going online to make purchases. Edge Cloud can be used to power applications like mixed reality mirrors in dressing rooms, which require constant interaction from customers, and it can help stores be more agile when it comes to introducing new features and updating software. In order to facilitate these kinds of interactions, low latency is a prerequisite for edge compute. Space efficiency is improved as a result as well. If a store has limited floor space and high rent, it may not want to invest in a large amount of specialized hardware to support these programs [29].

9.6   Retail Flow Analysis and Video Technology

In order to better understand the flow of customers, employees, users, and products throughout their facilities, businesses can employ Retail Flow Analysis and video technology. Actionable insights, such as where to stock high-value goods, where to stock goods that people will purchase on impulse, or where to place advertisements, can be gleaned from the analysis of these video feeds and used by the company. As a result, the network of cameras produces enormous amounts of data that must be sent back to and analyzed in the central cloud, a process that is both time-consuming and resource-intensive. Instead, footage can be analyzed and neutralized in a compliant and privacy-safe manner on the edge cloud located on the organization's premises using Edge computing.

9.7   Warehousing Real-Time, Tracking of Inventory

To keep up with this rising demand, logistics and retail companies need to know exactly what stocks are where in the supply chain at any given time, down to the level of individual products. To do this successfully, businesses need to gather extensive information on the whereabouts and condition of their goods in real time [29]. The use of edge computing would be useful in this case.

9.8   Robots

The use of robots in warehouses has the potential to increase productivity by lowering the error rate, thereby boosting customer satisfaction through both the reduction of errors and the subsequent reduction in lead times. Furthermore, with the help of IoT analytics and Edge computing, warehouse robots can independently move around the facility and retrieve the necessary supplies. When it comes to robot navigation, where autonomous vehicles must guide themselves through a warehouse while avoiding obstacles and other AGVs, real-time analytics are essential.

9.9   Advanced Predictive Maintenance

With the help of data analytics, sophisticated preventative maintenance can foresee when a machine will break down. In the event that a problem is anticipated, a notification can be sent to the appropriate people so that they can inspect the machine and fix any faulty parts or other problems. By doing so, we avoid a machine failure and its associated costs, thereby maximizing our return on investment. It is possible to implement predictive maintenance in vehicles by monitoring and relaying data on individual vehicle components' efficiency. These days, a digital device, typically a mobile computer, is installed in the car to perform this function.

9.10   Monitoring Driver Performance

Performance Tracking for Drivers Video analytics can be run at the edge of the network to keep tabs on a driver's actions, such as checking their speed to make sure they're not breaking the law and gauging how distracted they are while behind the wheel. Long-term analytics could be used, for instance, to train and improve safety, while the information itself could be used to alert drivers in the case of erratic or dangerous behavior. More data on the driver means more personalized brake schedules and maybe even better insurance coverage. Edge computing is useful for this application because it reduces the amount of data that must be transmitted to and stored in the cloud, concentrating the life stream at a single edge for subsequent analysis.

10   Conclusion

The concepts, architecture, essential technologies, security measures, and privacy safeguards that make up the edge computing paradigm are all laid out in detail in this article. Supporting the digital transformation of many industries and meeting the need for data diversification across a wide range of fields, edge computing provides data storage and computation at the network's edge in addition to nearby Internet intelligent services. Edge Computing is moving toward an open future, where it will converge with the use of data through AI and ML to generate actionable insights that benefit businesses and their customers. Someday, it will be treated like any other place, where programs can be installed without a hitch. Some have even suggested that edge computing will eventually replace cloud computing due to its enormous potential. Certainly, this developing field is expanding rapidly as sensors all over the Internet provide a deluge of data. Today's smart factories, smart grids, connected vehicles, and more are all made possible by edge computing. Edge computing has been propelled by IoT, but the technology fueled by 5G will have an even more profound impact in many industries beyond IoT.

References

1. Shi W, Pallis G, Xu Z (2019) Edge computing [scanning the issue]. Proc IEEE 107(8):1474–1481. https://doi.org/10.1109/JPROC.2019.2928287
2. What is edge computing? Everything you need to know. https://searchdatacenter.techtarget.com/definition/edge-computing. Accessed 08 July 2021
3. Edge Computing. https://nikku1234.github.io/2020-11-23-Edge-Computing/. Accessed 08 July 2021
4. Shi W, Cao J, Zhang Q, Li Y, Xu L (2016) Edge computing: vision and challenges. IEEE Internet Things J 3(5). https://doi.org/10.1109/JIOT.2016.2579198
5. Chen Y, Li Z, Yang B, Nai K, Li K (2020) A Stackelberg game approach to multiple resources allocation and pricing in mobile edge computing. Futur Gener Comput Syst 108:273–287. https://doi.org/10.1016/J.FUTURE.2020.02.045
6. Al-Turjman F. Edge computing: from hype to reality. [Online]. Available: https://books.google.com/books/about/Edge_Computing.html?id=hq14DwAAQBAJ. Accessed 08 July 2021
7. What is edge computing? | Cloudflare. https://www.cloudflare.com/vi-vn/learning/serverless/glossary/what-is-edge-computing/. Accessed 08 July 2021
8. IBM and Verizon Business Collaborate on 5G and AI Edge Computing Innovation | Blockchain News. https://blockchain.news/news/ibm-and-verizon-business-collaborate-on-5g-and-ai-edge-computing-innovation. Accessed 08 July 2021
9. Cao J, Zhang Q, Shi W (2018) Edge computing: a primer. Springer. https://doi.org/10.1007/978-3-030-02083-5
10. What is edge computing? Introduction to edge computing. https://stlpartners.com/edge-computing/what-is-edge-computing/. Accessed 08 July 2021
11. MEC 101: a primer for communications service providers | LinkedIn. https://www.linkedin.com/pulse/mec-101-primer-communications-service-providers-sandeep-sahu/. Accessed 08 July 2021
12. What is 5G everything you need to know Etechwire – Cute766. https://cute766.info/what-is-5g-everything-you-need-to-know-etechwire/. Accessed 08 July 2021
13. Top 5 benefits of edge computing. https://blog.wei.com/top-5-benefits-of-edge-computing. Accessed 08 July 2021
14. Al-Turjman F (2019) Edge computing: from hype to reality, p 188. [Online]. Available: http://link.springer.com/10.1007/978-3-319-99061-3. Accessed 08 July 2021
15. Mobile Edge Computing (MEC): what is mobile edge computing? https://stlpartners.com/edge-computing/mobile-edge-computing/. Accessed 08 July 2021
16. Shiyun tu. Accepted manuscript Edge cloud computing technologies for internet of things: a primer. [Online]. Available: https://www.academia.edu/35722241/Accepted_Manuscript_Edge_cloud_computing_technologies_for_internet_of_things_A_primer. Accessed 08 July 2021
17. Ksentini A, Frangoudis PA (2020) Toward slicing-enabled multi-access edge computing in 5G. IEEE Netw 34(2):99–105. https://doi.org/10.1109/MNET.001.1900261
18. Alizadeh M, Abolfazli S, Zamani M, Baaaharun S, Sakurai K (2016) Authentication in mobile cloud computing: a survey. J Netw Comput Appl 61:59–80. https://doi.org/10.1016/J.JNCA.2015.10.005

19. Karthikeyan P, Thangavel M. Applications of security, mobile, analytic and cloud (SMAC) technologies for effective information processing and management, p 300. [Online]. Available: https://books.google.com/books/about/Applications_of_Security_Mobile_Analytic.html?id=vdZdDwAAQBAJ. Accessed 08 July 2021
20. What are edge containers? https://www.stackpath.com/edge-academy/edge-containers. Accessed 09 July 2021
21. Shah SDA, Gregory MA, Li S, Fontes RDR (2020) SDN enhanced multi-access edge computing (MEC) for E2E mobility and QoS management. IEEE Access 8:77459–77469. https://doi.org/10.1109/ACCESS.2020.2990292
22. RStudio Tensorflow – RStudio Documentation. https://docs.rstudio.com/resources/tensorflow/. Accessed 09 July 2021
23. Architecting at the edge | IBM. https://www.ibm.com/cloud/blog/architecting-at-the-edge. Accessed 09 July 2021
24. Al-Turjman F. Security in IoT-enabled space
25. Xiao Y, Jia Y, Liu C, Cheng X, Yu J, Lv W (2019) Edge computing security: state of the art and challenges. Proc IEEE 107:1608–1631. https://doi.org/10.1109/JPROC.2019.2918437
26. 10 Edge computing use case examples – STL Partners. https://stlpartners.com/edge-computing/10-edge-computing-use-case-examples/. Accessed 09 July 2021
27. Al-Ansi A, Al-Ansi AM, Muthanna A, Elgendy IA, Koucheryavy A (2021) Survey on intelligence edge computing in 6G: characteristics, challenges, potential use cases, and market drivers. Futur Internet 13(5):118. https://doi.org/10.3390/FI13050118
28. Edge computing: why it matters | The Daily Star. https://www.thedailystar.net/bytes/news/edge-computing-why-it-matters-2031761. Accessed 08 July 2021
29. Edge use cases for retail, warehousing and logistics. https://stlpartners.com/edge-computing/edge-use-cases-for-retail-warehousing-and-logistics/. Accessed 09 July 2021

Big Data Analytics and Security Over the Cloud: Characteristics, Analytics, Integration and Security

Anjali Jain, Srishty Mittal, Apoorva Bhagat, and Deepak Kumar Sharma

1 Big Data: Introduction

Before building a comprehension of Big Data, let us recall what data is. Data refers to the information stored in the form of characters and symbols, which is stored and transmitted across different devices. Over time, as the number of devices increased, so did the volume of data. With huge amounts of data, there was a requirement for systems to store and analyse this data. As the name suggests, Big Data alludes to huge magnitude of information. The analysis of Big Data is called Big Data Analytics (BDA). Along with the size, Big Data encompasses much more comprehensive characteristics that we will discuss in this section. In this section, we discuss Big Data, its five Vs and why it is necessary in the present age.

1.1 What Is Big Data?

Big Data is a far-reaching concept that is not easy to define. However, it could be visualised as data which is huge in size, and grows exponentially with time in

A. Jain · A. Bhagat
Department of Instrumentation and Control Engineering, Netaji Subhas University of Technology, (Formerly Known as Netaji Subhas Institute of Technology), New Delhi, India

S. Mittal
Department of Electronics and Communications Engineering, Netaji Subhas University of Technology, (Formerly Known as Netaji Subhas Institute of Technology), New Delhi, India

D. K. Sharma (✉)
Department of Information Technology, Indira Gandhi Delhi Technical University for Women, New Delhi, India

© The Author(s), under exclusive license to Springer Nature Switzerland AG 2023
G. Srivastava et al. (eds.), *Security and Risk Analysis for Intelligent Edge Computing*, Advances in Information Security 103,
https://doi.org/10.1007/978-3-031-28150-1_2

both size and complexity. This data has high complexity, high uncorrelation and varies greatly in its structure and types of entities. This data is so complex that none of the traditional tools of storing structured data are competent for storing and analysing it. Subsequently, Big Data is the information that cannot be prepared utilising conventional tools [1].

An example of Big Data and its usage is in social media. It is the most used and more recognisable form of extremely great amounts of data, accounting to more than 500 TB of data injected into Facebook's database every day. Note that the types of data on social media is of large variety: text, images, videos and more. Another example is stock exchange data which changes rapidly. The fluctuation of stock prices creates multiple terabytes of data per day. The analysis of this data, and the decision of the investment portfolios of large firms is conducted using BDA every day.

The data that constitutes Big Data can be of various forms. The structure of the data could be of different extent, as discussed here:

- **Structured Data:** Any data which follows a consistent format is called structured data. Structured data can be put away or stored in a carefully fixed style. For example, data stored in relational databases are stored in the format of tables with columns as attributes. These databases provide ease of access and high consistency, but lack the flexibility of data storage.
- **Unstructured Data:** Data that does not follow any rigid format is called unstructured data. There are no fixed fields, types or representations. Any data whose format is not known is called unstructured data. An example of unstructured data is data that contains text, images and videos in a single instance. Unstructured data provide unlimited flexibility to the user, but processing and extracting useful information is very difficult as no known pattern is present.
- **Semi-Structured Data:** Data which comprises of both of the above discussed types is called semi-structured data. There is a loose format that is followed by the data, and data heterogeneity is provisioned. This data format combines the advantages of both the above-mentioned types. The data generated on the internet is mostly semi-structured data.

Big Data is data that could be any combination of the three data formats mentioned above. Thus, BDA provides large flexibility to the user in terms of the data design, while striving to maintain perfect consistency in the storage and transactions in the database. Usually, to reserve such large amounts of data, and access it effectively, we use data warehouses. Data warehouses are services and tools that enable transactions and storage of Big Data through OLAP and OLTP.

1.2   The 5 Vs of Big Data

Big Data characteristics are commonly represented by the "5 Vs of Big Data". These characteristics set it apart from traditional data. They are as discussed below and shown in Fig. 1.

Fig. 1 The 5 Vs of big data

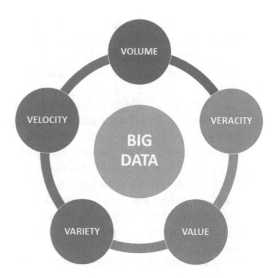

- Volume
- Variety
- Veracity
- Velocity
- Value

Let us now discuss each of these characteristics in detail.

1. **Volume**

    Volume corresponds to the large size of data. This is an essential characteristic for data to be classified as Big Data. The large, voluminous nature of data is the factor that expedited innovation of tools for storage and processing of such data, paving the way for entire fields in data science like data warehousing and data mining.

2. **Variety**

    The data in Big Data is highly heterogeneous with respect to its source, kind and nature. For a particular application, there can be multiple sources of data. For example, on an e-commerce website, the data of not just user activity on the website, but also the user's social media behavior, his or her location data, as well as the user's financial activity could be a few features that determine the product recommendations for the specific user. In terms of nature, the data could be structured, unstructured or a hybrid of the two formats as explained above. The type of the data is generally highly diverse, across various data types like text, image, PDF, videos and so on. Thus, Big Data has a very high variety.

3. **Veracity**

    The quality of data is represented by its veracity. Inconsistencies and anomalies in the data are caused due to the large volume and variety of the data. This could skew analysis, or create redundancies, hence using more storage. Therefore, Big Data must be clean and consistent. There must not be redundancies in the data. To achieve this, data preprocessing is employed, which performs data cleaning, data normalization and other such processes on the data to make it suitable for transmitting, storing and analyzing.

4. **Velocity**

The importance of data is time-bound. Certain data might be important only during an event, for example, the Super Bowl Sunday game or the Cricket World Cup finals. During these events, brands decide whether or not to promote their advertisements based on the viewership, sales and cost. After the event is over, that data has very little value for the brands. Similarly, in aircrafts, the speed of data transmission and analysis is needed to be as close as possible to real-time. This is because the trajectory depends on the data of GPS coordinates, weather, turbulence, air traffic and various other factors. Thus, BDA systems aim to achieve near real-time processing of data.

5. **Value**

The cost and effort put in analyzing such huge amounts of data must have a significant return on investment. There must be significant value generation from the data analysis being performed. If the data is no longer relevant, or the cost of analysis is more than the benefit then we say that this Big Data and analysis has minimum or no value. The aim of BDA should be to provide incremental profit to the organisation or the application.

These were the 5 Vs of Big Data, the characteristics that model the behavior of Big Data and distinguish it from traditional data. Next, we discuss in extensive detail why Big Data has turned out to be a necessity in today's day and age.

1.3   The Need for Big Data and Analytics

Since the advent of computer devices, data has only grown in volume and variety. Although the idea of storing and analysing large amounts of data is as old as data itself, the term 'Big Data' was authored very recently in 2005 by Robert Mougalas [2]. The evolution of Big Data can be used to understand how necessity gave birth to innovation. In the timeline given in Fig. 2, we consider the evolution of Big Data visualised in phases spanning over decades. Let us try to walk through this timeline to gain a better understanding.

The inception of World Wide Web (WWW) in 1989 as an information sharing project at European Organisation for Nuclear Research (CERN) sparked the idea to manage large amounts of data, especially data pertaining to the WWW.

In the early 2000s, the WWW gained popularity and offered unique use cases for the traditional databases. As the services provided by the internet increased, so did the expectations from the data and databases. Not only did the data increase in volume, but also types. Databases that held strictly formatted data till now were expected to manage different types of data like images, documents of various formats, and videos. This called for a modification to databases, and a possible

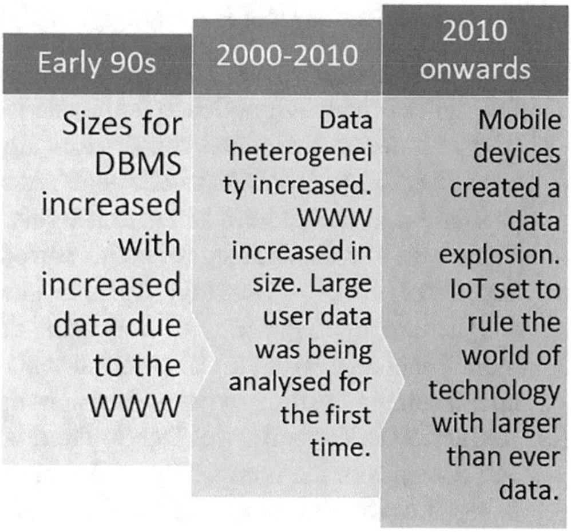

Fig. 2 Timeline for the evolution of Big Data

Early 90s	2000-2010	2010 onwards
Sizes for DBMS increased with increased data due to the WWW	Data heterogeneity increased. WWW increased in size. Large user data was being analysed for the first time.	Mobile devices created a data explosion. IoT set to rule the world of technology with larger than ever data.

transformation. Later on, as the internet saw the advent of global communication via social networking websites, the world became smaller while the data became larger [3]. The tech-giants of the time like eBay and Yahoo started to track user activity like click rates and geographical location. This gave a boom to the data analytics industry. The internet traffic increased, and with it, innovation in very large database management systems.

Of late, with advent of smart mobile devices, the current decade has been touted as the age of Internet of Things (IoT). We have seen an unprecedented explosion in the data that exists in the world. The overall data in the world in 2013 was estimated to be around 4.4 zettabytes [4]. By 2020, the data is set to reach a staggering 44 zettabytes. This amount of data is more than what can be processed by any existing technology that exists in the world. Therefore, Big Data is not a trend anymore, it is a necessity.

To summarise, we list down the reasons of the imperativeness of Big Data:

- Magnitudes of unstructured data created everyday.
- The count of devices is increasing with IoT, and so is the need for fast data storage and analysis.
- The need for data to move at a real-time speed in order to support the needs of the economy.

1.4 Advantages and Disadvantages of Big Data

We have already seen the need that is generated in the industry for Big Data systems. Let us now look at the advantages that are provided by the use of these systems [5]. Naturally, there are some disadvantages as well, which are also discussed.

The major advantages of Big Data and analytics are:

(a) **Improved productivity:** The ease of use and efficiency of Big Data tools allow for higher productivity of each employee in an organisation. Since the analysis of data is faster, individual contributors are able to perform more tasks in less time. In addition, Big Data can be analysed within the corporation to identify the strengths and weaknesses and gain insights into the organisation's behaviour to implement a wider strategy for better performance.

(b) **Insightful decision making:** By analysing much more data, we gain more insight into the behaviour of a system, market or application. The state-of-the-art data analysis and visualisation tools allow executives to cultivate better understanding, leading to better decision making.

(c) **Reduction in expenditure:** Due to the reduction in the time for each analysis, and decrease in the complexity of database systems, Big Data Analytics leads to significantly lower expenditure for organisations.

(d) **Faster results delivery:** Organisations have used BDA to expedite their deliveries to the market, by making use of Big Data's velocity.

(e) **Fraud Detection:** Advanced algorithms in BDA enable the system admin to automate the process of anomaly detection. This is exploited to detect fraudulent transactions and unusual behaviour in the network. This helps in ensuring the security of the network by tracking malicious attackers.

(f) **Induce innovation:** Since data is constantly evolving and customer needs changing, it nudges the industry to constantly innovate technology and tools to manage Big Data systems.

These were the advantages of technologies in Big Data [6]. However, due to the vastness of data and its heterogeneity, some disadvantages exist. These are:

(a) **Security Risks:** Due to the large-scale heterogeneity in data sources, there could be a vulnerability in the system for an external attack.

(b) **Skill deficiency:** Since this field is relatively new, this sector is not as established in terms of the skill set available in the industry.

(c) **Advanced hardware:** storing such large volumes of data with fast access poses unique hardware needs, which could drive up the cost as well.

(d) **Data quality:** The high variety of data could lead to lower quality. This can be minimised using data preprocessing techniques.

2 Analytics in Big Data

As data continues to grow bigger and bigger and decision making becomes data driven, there arises an inevitable need to churn that data to come up with meaningful insights which can potentially benefit many businesses and help them make better decisions and profits in the market. Since traditional data analytics is only capable of working on structured data, big data analytics emerged. This data can be any structured, unstructured, noisy or faulty data taken from a plethora of internal and external sources. Big data analytics successfully combats the shortcomings of traditional data analytics and is being extensively used in organisations to generate useful business insights.

2.1 What Is Analytics?

The process of analysing raw data using predictive and statistical techniques and drawing out meaningful inferences from it is known as analytics. Decision making is largely becoming data driven and business organisations have been using data analytics for a long time as it helps businesses make better decisions, study customer behaviours, perform market research and consolidate the most important business requirements [7]. Languages and tools like R, Python, Tableau, SAS and Microsoft Excel are extensively used for performing analytical tasks. Organisations employ data analysts, also called business analysts which add immense value to the firm by performing analytical tasks which greatly benefit the organisation.

2.1.1 Responsibilities of a Data Analyst

Following are the major responsibilities of a data analyst.

1. Acquire, process and analyse data
2. Derive valuable insights from data using statistical techniques
3. Use reporting tools to create comprehensive data reports
4. Identify patterns and trends in data
5. Establish and consolidate business requirements with inputs from fellow teams

2.2 The Link Between Analytics and Big Data

Big data analytics, which is made up of "big data" and "analytics", defines how analytical techniques are applied on big data. It is the process of uncovering hidden patterns and insights from large amounts of data. Big data analytics, alternatively called advanced analytics or statistical analysis, is based on the basic idea that the larger the data sample, the more accurate are the outcomes of the analysis. The evolution of powerful analytical platforms which are capable of handling big data have further popularised the use of big data analytics.

Table 1 Comparison between data analytics and big data analytics

Basis	Data analytics	Big data analytics
Suitable for large and complex datasets	No	Yes
Processing	Centralised	Distributed
Scalable	No	Yes
Type of data that can be worked with	Structured	All types (structured, semi-structured, unstructured, streaming data)

2.2.1 Comparison of Data Analytics and Big Data Analytics

Since the traditional data analysis techniques weren't designed keeping in mind the huge volumes of data which might not necessarily be structured in nature, big data analytics was introduced to deal with these issues and a scalable system was designed to work with large datasets which could contain structured, unstructured or even streaming data from enterprise level applications. A comparison between analytics and big data analytics has been presented in Table 1 [7].

2.2.2 Architecture of Big Data Analytics

The big data analytics architecture in Fig. 3 is divided into 4 sections, all of which are discussed below.

1. Collecting BIG data: Big data analytics works on the basic ideology which says that more the data the merrier, thus collecting this data is becoming of key importance. Data collection sources could be both internal and external including sensors, social media, data logs and the internet. To generate advanced business insights, data from enterprise and cloud applications can also accumulated.
2. Transforming big data: Post data accumulation from both structured and unstructured sources, this raw data needs to be processed so that it can be further utilised by big data platforms. This can either be done using data warehousing techniques like extract, transform and load (ETL) and the transformed data can be tracked using change data capture (CDC) [8].
3. Running data through big data platforms and tools: The transformed data is then sent as input to several big data platforms and tools as per the business requirements. Tools can be categorised into search tools, batch and streaming frameworks, SQL engines etc. and can be employed as and when required.
4. Applying big data analytics: The data is now ready to be used for big data analytics, query and reporting purposes, data mining and widely for statistical and predictive analysis [8, 9].

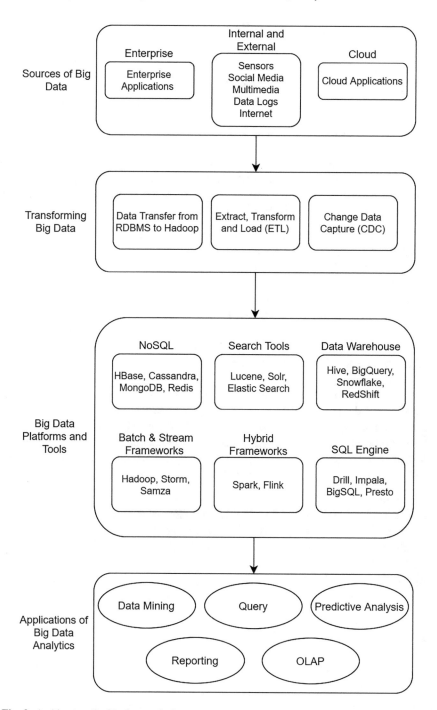

Fig. 3 Architecture for big data analytics

2.3   Benefits and Barriers of Big Data Analytics

Just like any other field of technology, big data analytics also comes with its pros and cons, a few of which have been discussed below.

2.3.1   Benefits of Big Data Analytics

Large number of business intelligence applications and nearly every customer facing application benefit from big data [10]. For firms which generate large volumes of data on a daily basis, big data analytics comes in handy and generates powerful insights. Some benefits of big data analytics are -

(a) Better decision making using accurate business insights
(b) Understanding customer behaviors better
(c) Credit risk and fraud detection
(d) Identification of opportunities in the ever-changing market
(e) Optimized planning and forecasting
(f) Optimum segmentation of consumer base

2.3.2   Barriers to Big Data Analytics

A survey was conducted in [10] to find the reasons why organisations refrained from adopting big data analytics in their firms. Lack of skilled labor, improper funding and unavailability of platforms capable of dealing with big data were found to be the key blockers. Big data analytics is a relatively new field and with requirement of specialized skills, that are different from the traditional business intelligence and data warehousing related tasks for which the systems are currently modelled for.

(a) High maintenance cost
(b) Difficulty in establishing the big data analytics system
(c) Incapability of handling heavy big data queries by the existing systems
(d) Existing data warehouses are modelled for Online Analytical Processing (OLAP) only
(e) Lack of skilled labor

3   Big Data Analytics in the Industry

3.1   Healthcare

Big Data can be understood as the digital storage of a large volume of various types of data. Healthcare is one data driven industry that has been constantly producing data. If stored and analysed properly, big data can be utilised in improving the available medical services [8], for example, medical costs, prevention of epidemic, detection and handling of avoidable diseases well in time and improving the quality of life in general. The following section defines the varied applications of Big Data in the healthcare sector [11].

3.1.1 Predictive Analysis

Predictive analysis is one of the most important business intelligence trends, it helps doctors to make fast and improved data-driven decisions. This proves to be helpful especially in cases with complex medical history and multiple sufferings [12]. As an example, a patient with diabetic trends can now be informed well in advance and take appropriate measures to reduce or eliminate the effect of the disease [13].

3.1.2 Smooth Hospital Administration

Storing patient data along with detailed information about hospital admission patterns, can help hospitals plan their logistics [14] and staff duties more effectively, so that there are always the right number of staff present at the hospital to smoothly manage the patient count. Additionally, this information can also be utilised to plan other medical supplies that must be made available at a particular time in the hospital.

3.1.3 Electronic Health Records (EHRs)

EHRs are digital records that contain detailed information about the medical history, diseases, medical prescriptions, laboratory test results, allergies and other hereditary abnormalities of a patient. Such records can be transferred through secure systems and be used to identify disease patterns, trigger timely alarms to suggest patients to undergo certain medications or tests [15]. EHRs also include an editable document that can be used by doctors, thereby eliminating any paperwork. Additionally, EHRs also reduce the number of visits to the hospital by a patient, as doctors and medical staff can now refer to this record to understand a patient's history and verify if a person has undergone the required tests for the treatment of a disease. This in turn, helps patients to reduce their medical costs [16].

3.1.4 Fraud Detection

Big Data analytics in healthcare can help prevent fraud by timely identification of fraudulent insurance claims [17] resulting in cost savings. This also proves helpful in prioritizing legitimate claims and enabling patients to get timely approvals for their claims along with the advantage of medical service providers being paid faster.

3.1.5   Medical Imaging

A large number of medical images are produced and studied by experts to identify meaningful patterns. However, storing and maintaining these images for several years is both cost and time consuming. Storing the images by converting them into numeric values based on the patterns established between image pixels using software algorithms built on top of Big Data Technologies can help [18] to store image data in a more compact and optimised manner and reduces the burden of a radiologist to study the images. Rather, experts can now analyse the meaningful patterns built by the algorithm and can process an increased number of images.

3.2   *Financial Services*

The finance industry generates huge amounts of structured and unstructured data that offers significant analytical opportunities and provides key decision-making insights. Big data together with data science, is used to provide quick finance analysis and make better informed decisions with respect to finance management. Big Data analytics provides investors with insights about investments with greater customer returns. Additionally, it has also facilitated algorithm trading that relies on huge volumes of historical data to build complex mathematical models to optimise the return on an investment. Insurance services and other financial institutions develop customer personalisation based on the customer's transaction history [19].

Financial services also leverage big data for risk management and verify creditworthiness of a customer. Increase in the number of financial transactions also increases the probability of frauds. Financial fraud detection is one of the most impactful use cases of big data in finance. Credit card frauds can now be detected with the use of advanced algorithms [20]. Tracing unusual trading patterns through big data can be used to minimise losses. Anomaly detection is much easier now with higher accuracy.

3.3   *Advertisement*

The goal of advertising and marketing is to educate, inform and persuade customers to buy products and services. The process of advertising can be optimised by defining the target audience and delivering the right customers who resonate with the idea of the product and services being sold. However discovering this target audience can be a challenging task without data analytics. Big data and appropriate data mining facilitate the capture of user behaviour and establish customer interest patterns that might have otherwise been overlooked [21]. For example, OTT media services mine users' watch history to make a decision on the kind of content to be hosted on their platform. Target advertising ensures high profits for businesses as well as enhances user experience.

Unstructured big data can facilitate customer acquisition and retention. By leveraging data about a user's likes and dislikes, subscriptions, companies can now devise tailored product and promotional strategies that cater to the requirements of the consumer. Big data in advertising can also optimise media scheduling. With the ability to scale big data, this data can now be analysed at a granular level eg: ZIP code level. This information can be utilised to streamline both the time and content of advertisements to be displayed to the right audience in a particular region. In addition to this, product development can be triggered by observing the response to targeted advertisements to maximise profits. Thus, big data is vital in informing the constant shift in customer preferences and it creates an ecosystem where customer experience is upgraded.

3.4   Case Study

E-commerce services use Big Data collected from their customers with their browsing history in order to optimise its recommendation engine. The more the e-commerce service knows about its customers, the better it can predict the customer's interests. Once a seller has identified the interests, the process of streamlining the recommendations to provide to its customers becomes easier. This also enables the retailers to persuade their customers to buy a product in a much more efficient and compact manner that satisfies the goal of the e-commerce service as well as enhances customer experience.

There are several techniques through which the process of recommendations can be implemented. One of these techniques is collaborative filtering. Collaborative filtering can also be referred to as profile matching. This means that an e-commerce service builds a picture of who you are, and then decides what to recommend to you based on what other profiles similar to yours have purchased or shown interest in. Thus, the e-commerce service builds a "360-degree view" of each of its customers. Apart from profile matching, these services also implement product matching and associations to suggest to its customers the set of products that should be bought together. For example, a customer who searches for a travel pillow shall also be recommended to buy an eye mask with it, thus presenting to the customer valid associations, and eventually the customer ends up buying both the products. This results in a win-win situation in terms of both customer fulfilment and retailer's profit maximisation.

Questions
1. Based on the above reading, how do you think Big Data facilitates the functioning of e-commerce services?
2. What are the data attributes (customer data and product data) that will be required to implement collaborative filtering and product associations?
3. What are the possible Big Data challenges that an e-commerce service can face?

4   Big Data Technologies

Data nowadays is being generated at an exponential rate and there arose a dire need for platforms which should store and process such large volumes of data in the early 1990s. With the popularization of Big Data, frameworks like Hadoop, Spark and Flink came up, and soon came the NoSQL databases which could efficiently store both semi-structured and unstructured data. Existence of such platforms led to even more innovation and growth in the field of big data and some of these technologies have been discussed below.

4.1   NoSQL Databases

"Not Only SQL" databases, also called NoSQL databases are non-relational databases which are capable of storing large volumes of both semi-structured and unstructured data. NoSQL follows the CAP theorem (in Fig. 4) i.e. Consistency, Availability and Partition tolerance, which allows a database to meet only two out of three needs of the database at a time, i.e. the database can either meet the Consistency and Availability (CA) requirements or it can meet the Consistency and Partition tolerance (CP) or it can meet the Availability and Partition tolerance (AP) requirements at a time.

4.1.1   Features of NoSQL

A NoSQL database is characterised by the following features [22]:

Non-Relational

Unlike relational databases in which the data is normalised, NoSQL stores data in a denormalized manner. The data can be stored in a flexible manner which is most suitable for its structure and type. It doesn't necessarily needs to be modelled in a tabular structure of rows and columns as in a relational database.

Fig. 4 CAP theorem

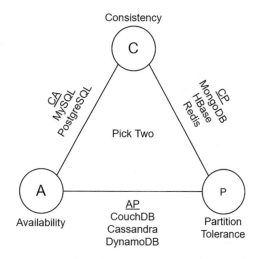

Schema Less

NoSQL has a flexible schema and does not require a fixed data structure in order to store data. The data is modelled in the data structure where it fits the best and is ideal for storing semi structured and unstructured data.

No Joins

Joins like in SQL, is not supported in NoSQL since the latter is a distributed database and would require a large amount of computational and operational overhead during key lookups in the database.

Highly Scalable

NoSQL databases scale out horizontally by adding more robust servers in order to handle increased traffic as compared to vertical scaling which involves adding more power to an existing node/machine.

4.1.2 Types of NoSQL Databases

NoSQL databases can be divided into four types namely key-value based, graph based, document based and column based [22] and have been illustrated in Fig. 5.

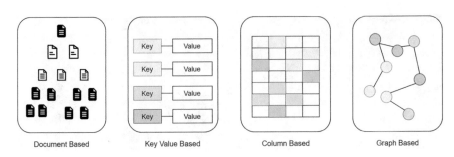

Fig. 5 Types of NoSQL databases

Key-Value Based

These are the simplest kind of NoSQL data stores in which data is stored in the form of key-value pairs as in a hash table making it scalable and provides a very high query-response time. Keys are bound to be unique and the value can belong to any data type. Some examples of key-value based databases include Redis, Memcached and Riak.

Column Based

To reduce the overhead of scanning through all the columns in traditional row-based databases, column-based databases store each column separately, thus allowing quicker retrieval of data when information is required only from a small number of columns. Examples include HBase, Cassandra, Hypertable and Google BigTable.

Graph Based

These databases store data in a graph-like data structure. It is capable of storing entity relations and is made up of nodes and edges where the nodes signify the entities and the edges signify the relationship between two entities. Examples include Neo4J, OrientDB and FlockDB.

Document Based

These databases are an extension of the key-value based databases with the difference that here the value is a document which is stored in the XML or JSON format. Examples include MongoDB and CouchDB.

4.2 Hadoop

Soon after Google developed the Google File System (GFS) in the early 2000s for data intensive applications, the MapReduce framework was developed in order to facilitate parallel computations [23]. Apache Hadoop is essentially an implementation of the MapReduce framework [24]. Hadoop allows distributed storage and processing of large volumes of data across clusters of computers. It is highly scalable and is capable of performing error detection and recovery on its own. Hadoop 3.3.0, which is Hadoop's latest release contains five modules which are Common, Hadoop Distributed File System (HDFS), Yarn, MapReduce and Ozone [25]. The characteristics of these modules have been discussed below.

4.2.1 Hadoop Common

It includes the common utilities and libraries that are needed for the functioning of other Hadoop modules. It also includes some Java Archive (JAR) files and scripts which are needed to start Hadoop.

4.2.2 Hadoop Distributed File System

It is a distributed and scalable file system component for Hadoop. It is run on commodity hardware and is capable of storing large amounts of data. It follows a master-slave architecture. HDFS has been discussed in detail in Sect. 4.3.

4.2.3 Hadoop YARN

As the name suggests, Yet Another Resource Negotiator or YARN is a framework for resource management and job scheduling on Hadoop's computation platform. It is one of the core components of Hadoop which can dynamically allocate resources to applications as and when needed. As compared to MapReduce's static resource allocation approach, YARN has been designed to improve utilisation of resources and performance through dynamic resource allocation [26].

4.2.4 Hadoop MapReduce

MapReduce allows parallel and distributed processing in Hadoop clusters. As the name implies, it consists of two tasks which are Map and Reduce and as per the sequence, the reduce task is always performed after the map task. MapReduce has mostly been used for batch jobs so far and happens to be very scalable. More about MapReduce has been discussed in Sect. 4.4 [27].

4.2.5   Hadoop Ozone

Ozone is a relatively new addition to the Hadoop framework and is essentially a distributed object store for Hadoop. It is built on top of Hadoop Distributed Data Store (HDDS). Its high scalability and availability allow it to operate efficiently in containerized environments like Kubernetes and YARN [28].

4.3   Hadoop Distributed File System (HDFS)

As discussed briefly in Sect. 4.2.2, HDFS is the distributed storage component for Hadoop. It runs on commodity hardware and follows a master slave architecture. It uses the TCP/IP protocol for communication and is very robust in the sense that it does efficient fault detection and recovery.

4.3.1   Objectives of HDFS

HDFS has been evolving over time and these are the goals which HDFS eventually aims to achieve as it gradually improves.

(a) Fault detection and recovery in the case of namenode or datanode failures.
(b) Ability to deal with large data sets and files.
(c) Accessibility to applications through APIs.
(d) Ability to recover files from the trash directory within a stipulated time.
(e) Store large volumes of data reliably using data replication.
(f) Portability across external software and hardware platforms.

4.3.2   HDFS Architecture

HDFS is based on a master slave architecture and is made up of a single namenode and a cluster of datanodes along with HDFS clients [29]. The architecture of HDFS has been illustrated in Fig. 6 and its components are discussed in detail below.

(a) NameNode: Each HDFS cluster contains a single namenode which acts as the master in the master-slave architecture. It is essentially a piece of software which is run on commodity hardware and runs the GNU/Linux operating system inside. Namenodes regulate how clients access the files. File operations including opening and closing of files and directories are also dealt with by the namenode.
(b) DataNode: For every node in an HDFS cluster, usually a single datanode is associated with it and is responsible for managing the storage of the node they are associated with. Similar to namenodes, it is also a software which runs on commodity hardware and runs the GNU/Linux operating system inside [30].

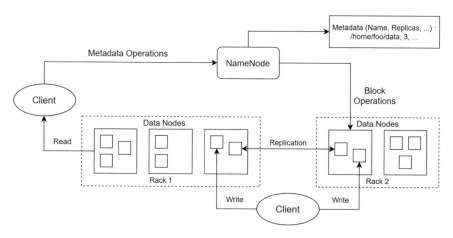

Fig. 6 Architecture of HDFS

Datanodes themselves contain some blocks which are obtained by splitting the files to be stored. The master node (namenode) instructs the datanodes to send block reports and even to register again or shutdown [24, 29].

(c) HDFS Client: Clients do the read/write operations from the datanodes. The client sets up a TCP connection with the namenode and the namenode uses the datanode protocol to connect with the datanode. It is through Remote Procedure Call (RPC) that the HDFS clients and the datanodes actually communicate [29].

4.4 MapReduce

As discussed briefly in Sect. 4.2.4, the MapReduce framework allows distributed computing of large volumes of data parallely on a large cluster of nodes. It consists of two major tasks which are the Map task and the Reduce task. A MapReduce task splits the incoming data into some independent blocks which are then parallely processed by the Map stage. The output from the Map stage is then sent as input to the Reduce stage. Incase of failures, the MapReduce framework deals with rescheduling and re-executing the tasks.

4.4.1 Phases of MapReduce

In order to implement the MapReduce job, applications usually implement the Mapper and the Reducer interfaces for the Map and Reduce tasks respectively. The Reducer is divided into the Shuffle and Sort and the Reduce phases [31]. Before the data is fed to the mapper, it is split into smaller chunks (or input splits) which are sent to the mapper one at a time. The map and reduce task have been discussed below.

Map Task

Each map task accepts an input split as input from the HDFS and transforms them into key-value pairs after applying the mapping function to each split. Then after the map function has been run on each split, the output of the mapper is registered with the TaskTracker which then informs the JobTracker that the map task has been completed.

Reduce Task

The two phases of the reduce task are -

(a) Shuffle and Sort phase: Output from the map task is sent as input to the shuffle and sort phase. It is responsible for combining the identical key-value pairs obtained from each mapper's output and combining them together [27]. The output of the shuffle and sort phase is partitions of key-value pairs where each partition contains no two different pairs.
(b) Reduce phase: This phase combines all the key-value pairs in a single partition to one pair and changes the value of the combined key-value pair to the number of times the key-value pair appears in the partition in the shuffle and sort phase [27]. Output from the reduced phase is rendered as the final output.

5 Big Data Analytics and the Cloud

Big Data Analytics and cloud computing are two new technologies serving different functions. However, when used together, they provide state-of-the-art solutions for business needs. In an attempt to understand the roles and relationships between these, let us first develop an understanding of cloud computing. This is followed by understanding the relationship between the two, and how they come together. Finally, we talk about cloud services and their uses in big data analytics.

5.1 Gentle Introduction to Cloud Computing

Cloud computing has been enabling industries to grow by leaps and bounds in terms of scale by providing complex infrastructure to even small businesses over the internet. Due to cloud computing, the large and expensive IT infrastructure has now moved to the internet. Cloud computing can be defined as a service which provides on-demand services for the user's computing needs: starting from storage, computing, analytics and so on [32]. These services are usually rented or leased from cloud service providers, with a pay-as-you-go model. This refers to paying for only those services that you use, for as long as you use them. This ensures minimal costs and higher trust generation amongst the users. Due to these benefits, cloud has been embraced by almost all industries for their storage and computing needs. What is in it for the cloud service providers? Since the costs of cloud services are minimal, one wonders what the cloud service providers stand to gain from this deal. The answer to that is scale: cloud service providers gain through economies of scale. When the number of users for the services increases, the cloud services providers make profits [6].

Various services including networking, processing power, storage and so on are provided by cloud computing. There are several examples for displaying the usage of cloud computing around us today. The most prominent example from our everyday usage is the cloud storage that is available for our mobile devices, where users can backup their data. This is an example of storage over the cloud. Another example is the online streaming OTT platforms. These websites use the cloud processing services to stream the data across the world.

The importance of cloud computing can be understood using the statistics of global spending on cloud services. In 2016, the global spending was valued at approximately 20 billion USD. By 2020, this figure will increase to around 50 billion USD. As we can clearly see, there is a huge investment in cloud services. In addition, the growth in cloud services engagement is much faster than was estimated. Therefore, cloud computing is the new way to use the internet, software and platforms.

5.2 Relationship Between Big Data and the Cloud

With an understanding of cloud computing, we understand that it provides infrastructure for large scale storage and processing. Where cloud represents infrastructure, Big Data represents content. There is a need for services which are competent in storing, managing and analysing this data securely and efficiently, with minimal costs due to the large volume of big data. The natural association between Big Data and cloud computing becomes imminent- cloud technologies are used for Big Data management [33]. In this manner, two technologies from different spheres are used synergistically to provide highly functional, secure and efficient solutions for the industry. In the following section, let us explore the connection between Big Data and cloud computing.

5.2.1 Synergy Between Big Data and Cloud Computing

The highly beneficial relationship between Big Data and Cloud Computing manifests in the following ways:

1. Cloud computing provides high complexity infrastructure for the storage of big data. The incessant increase in the size of big data poses concerns for data engineers, as there are no traditional tools and technologies to deal with data of this magnitude. Therefore, cloud computing is employed to store Big Data, as cloud technologies are scalable and adapt to store high volumes of data.
2. As cloud computing is a form of distributed computing, it is spread in geographically spread out areas. This property of cloud computing is levied for the use of Big Data as a way of having highly diverse sources of data, gathered from all over the world over the cloud. High variability to Big Data is provided by cloud computing, thus making it more comprehensive.
3. Data visualization is a powerful way of gaining and presenting the patterns, insights and important information derived from the large clusters of structured and unstructured data that forms big data [34]. Many cloud services provide solutions for Big Data transactions, analysis and management. Thus, cloud computing provides the tools required to cultivate value from big data; the use case of these tools is provided by Big Data.
4. Since both these tools are relatively new, they are in the initial stages of developing their niche in the economy. As Big Data grows rapidly, cloud computing technologies need to be progressively improved and new technologies need to be innovated [35]. There is huge potential in both these technologies for innovation and research. Thus, both of these technologies incrementally advance each other.
5. Due to the pay-per-use model of payment for cloud services, the processing of big data can be done in a cost effective manner.

5.3   Cloud Services for Big Data Analytics

In the previous section, we talked about the synergistic relationship between Big Data and cloud computing. We discussed that cloud computing provides various services for the storage and management of big data. In this section, we talk about specific services provided by the cloud, and their uses in big data.

The services offered by cloud computing are categorized below:

1. **Platform as a Service (PaaS):** Cloud vendors provide frameworks and tools which can be used by developers to customize according to their use and build applications. For this delivery, the cloud services provider manages all the resources and middleware, for example operating systems, software updates,

storage and so on. This enables the developers to build their application without having to worry about the extra dependencies [36]. There are several PaaS that are used for big data applications, like Microsoft Azure, Amazon Web Services and so on.

2. **Software as a Service (SaaS):** SaaS is a service similar to PaaS, where instead of platforms, software is provided to the end user to be used via the web, without downloading. The hosting and deployment of the software is taken care of by the provider of the cloud service. All that is left for the end user to do is to simply customise the application and use it directly via the web. There exist many SaaS for big data analytics, including Tableau, Kafka and many more.

3. **Infrastructure as a Service (IaaS):** Cloud service providers serve IaaS in the form of servers, networks and storage. IaaS can be used with complete customisation by the user, just as a traditional data centre. This service is provided in the form of virtualisation of the hardware. For BDA, there are various infrastructure needs that are solved by IaaS like XenonStack [37], Azure and so on.

6   Security Over Cloud

6.1   Security Concerns

The cloud provides a variety of services and flexibility to users in multiple ways. However, the cloud cannot be considered to be entirely secure from attacks and failures. Because of the fundamental nature of availability of services over the internet, there are several vulnerabilities associated with the cloud. Security concerns such as the ones mentioned in Fig. 7, are some of the major risks that prevent the cloud from being entirely secure.

The following section describes some of the security issues in detail and discusses certain approaches to protect data and secure the cloud vulnerabilities.

Fig. 7 Secure concerns in cloud computing

6.1.1 Data Breaches

Large enterprises and businesses leverage the cloud storage service to store structured and unstructured data on the cloud. As discussed earlier, this gives businesses various advantages like reduced costs, higher flexibility and collaboration options. However, storing sensitive data online, in a shared environment, has its own concerns; one of the major concerns being that of data breach. Data breach is simply a leak of private and confidential information to untrusted parties, whether on purpose or by mistake [38]. The reasons behind a data breach can be many, ranging from personal benefit, organized crime association or political activism. The large amounts of data generated through various IoT devices acts as an incentive for data breach [39–42].

One of the major reasons for data breaches over clouds occurs due to human error while setting up cloud security services. The cloud security controls are often complicated and are easily overlooked. Improperly set up servers and misconfigurations act as an easy target for the hackers thereby exposing the system vulnerabilities and resulting in easy access to sensitive information. These misconfigurations can be due to lack of logging, improper permission controls, unrestricted access to services, lack of proper data encryption and many more. The cost and other consequences related to cloud data breach is extremely severe. If databases such as those that store the citizen registry are accidentally left for public access, it compromises sensitive data such as personal information, name, address etc. of millions of people and exposes it to multiple unauthorised activities. Information regarding intelligence and defence services, if breached, can result in a very severe threat to the world.

6.1.2 Account Hijacking

With growing number of user accounts and subscriptions, and organisations shifting their implementations on the cloud, account hijacking poses a big issue in the current times. Cloud account hijacking is a process through which cloud account credentials are stolen or hijacked that essentially results in an identity theft. In several cases, email accounts are linked to various other platforms such as financial transactions services and account hijacking in such cases have extreme implications. When an attacker acquires cloud account login information, it can be used to access sensitive data or carry out malicious activities such as falsify content, manipulate credentials to remove hijack suspicions or script bugs in the existing systems [43].

Account hijacking can be performed through phishing, multiple attempts at password guessing or sending fake emails. Cloud services make use of user tokens to eliminate the need for a user login every time an update or sync is performed, however this acts as a vulnerability in terms of account hijacking. Lack of strong authentication services, multi-factor authentication processes based on dynamic passwords and unrestricted access to all IP addresses to access cloud applications are some of the many reasons that facilitate account hijacking and must be taken into account to prevent the same.

6.1.3 Insider Threat

Insider Threat refers to the threats posed by authorised identities within an organisation to misuse information on cloud applications. Employees might use their access to gain sensitive data either with malicious intent or by accident. Such a threat compromises an organisation's confidentiality, integrity and data availability. For instance, if an ex-employee's account access is not disabled on time, he/she can still access the services over the internet, with very reduced credibility and accountability [44]. Such an access can be used to keep a copy of sensitive information and eventually can be sold to a competitor which can result in grave business losses. Insider threats can be reduced by limiting access within an organisation while also actively monitoring activities to detect abnormal behaviour eg. spike in the number of data downloads.

6.1.4 Denial of Service Attack

In the types of security risks discussed above, the attack is intended at hacking the systems to get access to sensitive information. However, Denial of Service (DoS) is a cyber attack which doesn't aim to hijack sensitive data, rather it is associated with making a service unavailable. This means, in most cases, DoS attacks will not result in a security breach, rather, it will disrupt legitimate users from accessing services and applications hosted on the cloud. As shown in Fig. 8, DoS attack is executed by a malicious entity who floods the target machine or cloud server with surplus and redundant requests with the aim of overloading the system, as a result of which these systems are unable to fulfil legitimate requests, hence denying services to the latter [45].

A DoS attack can be dealt with by identifying the source of these multiple requests, and blocking any requests from the identified malicious source. However, a variation of the DoS attack, is the Distributed DoS (DDoS) attack, wherein these superfluous requests are made though distributed sources ie. multiple systems now flood the cloud instead of just one as in the DoS attack. Detection and Prevention of a DDoS attack, thus, becomes difficult because of its distributed nature.

6.1.5 Malware Injection Attack

An increased number of dynamic web pages, web services and web applications are being hosted on the cloud due to the flexible and scalable nature of the latter. Malware injection attack is a type of web-based attack, wherein attackers inject malicious programs into a web based application by using the weak links in the security. These malicious applications can be executed to perform a variety of actions such as data manipulation, eavesdropping or data thefts. Cloud computing is at risk from malware injection attacks because these applications mimic to be legitimate applications [46]. Hackers design malicious software applications,

Fig. 8 Denial of Service attack

programs or virtual machine instances and inject it into cloud services. Once inside the cloud network, this manipulated service injected by the hacker now performs the intended manipulations based on the code embedded in it.

6.2 Countermeasures for Cloud Security Issues

As discussed above, the cloud layer is not completely immune to security risks and attacks. However, implementing some countermeasures can help to increase the security of the cloud. Advanced security technologies combined with strict security policies can enable cloud service providers and users to enhance secure cloud services. Getting access to cloud services is an easy process that can be achieved through credit card transactions. However, such a weak registration system without other strict policies and terms of usage, allows hackers to get access to cloud services through credit card frauds or other fraudulent methods. With the use and implementation of stringent rules and regulations, network administrations and cloud providers can better manage the cloud efficiently. Terminating service instances on violation of user policies, can help reduce and eliminate malicious activities [47].

In order to eliminate data breaches and data loss, it is essential that valid data protection schemes are implemented and advanced encryption tools are used for storing sensitive data. Proper and continuous monitoring of the cloud is required to identify anomalous user behaviour and patterns to further prevent the origin of malicious activities through account blocking or other warnings, as described in the cloud service policy. Several machine learning based tools can be employed to detect actions intended towards data loss or cloud service disruptions.

To prevent attacks such as the DoS, it is important that process access management is put into effect. It is essential to monitor the network traffic and restrict access to data and services through security techniques. Intrusion Detection Systems and firewalls are some tools that can be used to restrict unauthorised parties from gaining access to sensitive information. To monitor the use of cloud software tools and the related data, there are tools like eXtensible Access Control Markup Language (XACML) and Security Assertion Markup Language (SAML). Periodic checks for authenticity through token validations or authentication through virtual and physical keys can ensure legitimacy of the users [48].

With the ever growing amount of data and on-demand service requirements, it is essential that the continuous technological developments make use of cloud computing. While cloud provides various services, it also suffers from some key challenges that cannot be ignored. However, to ensure seamless service, security flaws need to be continuously detected and eliminated through dedicated mechanisms by both the providers and users of cloud. The shared responsibility of the cloud between users and providers, must be carefully understood in order to develop detection and prevention techniques for data losses and security attacks.

7 Challenges and Research Areas

Big data analytics and security concerns over cloud computing have continued to evolve over the years and there still exist some challenges to them which are yet to be completely resolved. Research is being conducted to meet the recurring challenges and the open challenges in the fields of big data and cloud computing have been discussed below.

7.1 Open Challenges in Big Data

Although the advent of big data and BDA has proved to be a major blessing for many business organisations, there still exist some challenges to its adoption in many firms. Three of the five Vs of big data: variety, volume and velocity continue to pose challenges for organisations which can't cope up with the increasing dimensionality of the data and lack of storage space to store the data [10]. Surely there are huge volumes of data being generated each second but it can not be denied that there exists a lot of inconsistency and uncertainty in that data. The uncertainty lies not just in the mode of collecting data but also depends on the data source, the completeness of data and if the data is ambiguous or holds any inaccuracies. Inefficiency and poor performance of data visualisation tools continues to prove to be a hindrance in presenting insights obtained from big data analytics [49].

There also exists a dearth of tools which can efficiently retrieve meaningful information from big data along with its suitable management and representation. This highly hampers the knowledge discovery process. The high degree of computation power required to process big data continues to remain an open challenge for big data and research still continues to develop techniques to minimise memory requirements and computational demands for handling big data. Coming to big data analytics, uncertainties do exist in how big data analytics is performed using artificial intelligence. Incomplete and noisy training data and unlabelled data in the case of supervised learning contribute to big uncertainties in machine learning tasks and lead to inconsistent classification [50]. Maintaining information security is another challenge with big data analytics and measures to secure big data are being constantly developed.

7.2   Open Challenges in Cloud

It is now known that cloud computing is revolutionising the IT industry. However, there still exist some challenges as the adoption of cloud grows exponentially. Cloud security risks have been one of the top concerns for most businesses. The inability to see the exact location of data storage and processing increases risks encountered during the management or implementation of services on the cloud [51]. Secondly, even though migrating services to the cloud eliminates the need to purchase and maintain the hardware, it is essential to point out that it is not always possible to predict the costs and quantities required to scale your services. Thus, keeping a check on cloud costs becomes difficult. Often, it becomes tough for businesses to cope with the pace at which the cloud computing tools are advancing. This leaves a company or an enterprise with a lack of expertise or resources to manage cloud services [52].

Another major issue with migrating to the cloud is that of compliance. Any company that plans to shift its internal local storage to the cloud must follow industry standards and laws and ensure proper compliance regulations. Once the services have been migrated to the cloud, the performance of these services now depends on both, the business owner as well as the cloud provider [53]. If the cloud goes down, the services hosted on the cloud go down with it as well. It is essential for businesses to choose the right providers that have well implemented alarm notification systems in place. Companies planning to shift to build a private cloud face multiple additional issues with respect to automation and implementing privacy. Moreover, migrating fully functional services to the cloud has its own set of issues with respect to troubleshooting, slow data migrations, application downtime and much more [54].

7.3 Future Scope

In the coming future, the cloud will play a key role in how we manage big data. Keeping in mind the exponential rate with which big data is growing, only the cloud will be able to store the high volumes of data. Processing big data will get easier with machine learning. Machine learning will be used not just to generate relevant insights from data but also for data visualisation [55]. With the development of more advanced machine learning algorithms, the ability of computers to learn from data will greatly increase. The future of big data analytics will bring a new face to healthcare, finance, e-commerce and many other industries. Although big data might continue to pose some challenges because of its overwhelming size, it is bound to add significant value to all businesses.

The technological advancements provided through cloud computing with respect to storage and processing is bound to act as a great advantage for businesses in the future. With appropriate security measures in place, the cloud technology can provide highly personalised features to satisfy all kinds of business needs. For improved customer service, it is essential to choose the right cloud application or tool out of the ever growing set of cloud services being made available by cloud providers [56]. The microservices based architecture of the cloud will result in increased flexibility in the future. Automations tools for cloud management and multi-cloud based services are some of the innovations that promise a new revolutionised era in the field of cloud computing.

8 Conclusion

This chapter heavily focuses on Big Data and Analytics (BDA) along with its relationship with cloud computing. We began with an extensive discussion on what Big Data is. We developed an understanding for the 5 Vs of big data. The 5 Vs characterize Big Data and distinguish them from the traditional data. We understood the need for big data in today's world. Also, there was a discussion on the advantages and concerns regarding big data. The value of Big Data is mainly derived from the analytics of this data. We studied BDA in detail, including the related benefits and barriers. BDA is used extensively in various industries, including healthcare, finance, advertisement and so on. Through the case study on e-commerce services, we understood the real-word applications of BDA. For processing, storing and managing Big Data, various tools and technologies are brought to use. We talked about the technologies used in BDA, like Hadoop, HDFS and MapReduce. Since the data in Big Data is highly heterogeneous, traditional RDBMS is not competent for its management. Therefore, we studied about NoSQL databases.

After studying the comprehensive details about big data, Big Data and Analytics (BDA) and the related tools and technologies, we explore the connection of big data with cloud computing. Big Data and Cloud computing are both different technologies, but with a synergistic relation when used together. They provide highly functional, scalable, cost effective business solutions. We began with a gentle introduction to cloud computing, and eventually proceeded to explore the relation between big data and cloud computing. We then talked about cloud services and how they are used for BDA. Finally, we discussed the security concerns of the cloud. We explored possible attacks on cloud services, which are not uncommon in the context of cloud applications. We then discussed in brief the countermeasures that can be taken to make the cloud more secure.

Through this chapter, we explore the unique and highly beneficial relationship between Big Data and Cloud computing. This was done using a holistic approach based on thorough understanding of both the technologies. By addressing all the related aspects of Big Data, we ensured a sound understanding of the topic. Through the real-world case study, we provided a notable improvement in the learning process by establishing the relevance of the matter explained. This also provides the benefit of facilitating the imagination of the manifestations of Big Data in the industry, highlighting the advantages and concerns in a specific use case.

References

1. Zikopoulos P, Eaton C, IBM. (2011) Understanding big data: analytics for enterprise class Hadoop and streaming data, 1st edn. McGraw-Hill Osborne Media
2. Devgan M, Sharma DK (2019) Large-scale MMBD management and retrieval. In: Tanwar S, Tyagi S, Kumar N (eds) Multimedia big data computing for IoT applications. Springer, Singapore, pp 247–267
3. Chen M, Mao S, Liu Y (2014) Big data: a survey. Mobile Netw Appl 19:171–209
4. Gandomi A, Haider M (2015) Beyond the hype: big data concepts, methods, and analytics. Int J Inf Manag 35:137–144
5. Devgan M, Sharma DK (2019) MMBD sharing on data analytics platform. In: Tanwar S, Tyagi S, Kumar N (eds) Multimedia big data computing for IoT applications. Springer, Singapore, pp 343–366
6. George G, Haas M, Pentland A (2014) Big data and management. Acad Manag J 57:321–326. https://doi.org/10.5465/amj.2014.4002
7. Tsai C-W, Lai C-F, Chao H-C, Vasilakos AV (2015) Big data analytics: a survey. J Big Data 2(1):1–32
8. Raghupathi W, Raghupathi V (2014) Big data analytics in healthcare: promise and potential. Health Inf Sci Syst 2(1):3
9. Chan JO (2013) An architecture for big data analytics. Commun IIMA 13(2):1
10. Russom P (2011) Big data analytics. TDWI Best Pract Rep Fourth Q 19(4):1–34
11. Asri H, Mousannif H, Al Moatassime H, Noel T (2015) Big data in healthcare: challenges and opportunities. In: 2015 international conference on cloud technologies and applications (CloudTech). IEEE, pp 1–7
12. Wang Y, Kung LA, Byrd TA (2018) Big data analytics: understanding its capabilities and potential benefits for healthcare organizations. Technol Forecast Soc Chang 126:3–13
13. Eswari T, Sampath P, Lavanya S (2015) Predictive methodology for diabetic data analysis in big data. Procedia Comput Sci 50:203–208

14. Fihn SD, Francis J, Clancy C, Nielson C, Nelson K, Rumsfeld J, Cullen T, Bates J, Graham GL (2014) Insights from advanced analytics at the Veterans Health Administration. Health Aff 33(7):1203–1211

15. Bates DW, Saria S, Ohno-Machado L, Shah A, Escobar G (2014) Big data in health care: using analytics to identify and manage high-risk and high-cost patients. Health Aff 33(7):1123–1131

16. Ross MK, Wei W, Ohno-Machado L (2014) "Big data" and the electronic health record. Yearb Med Inform 9(1):97–104

17. Konasani V, Biswas MM, Koleth PK (2012) Healthcare fraud management using big data analytics. An unpublished report by Trendwise Analytics, Bangalore, India

18. Morris MA, Saboury B, Burkett B, Gao J, Siegel EL (2018) Reinventing radiology: big data and the future of medical imaging. J Thorac Imaging 33(1):4–16

19. Hussain K, Prieto E (2016) Big data in the finance and insurance sectors. In: New horizons for a data-driven economy. Springer, Cham, pp 209–223

20. Kamaruddin SK, Ravi V (2016) Credit card fraud detection using big data analytics: use of PSOAANN based one-class classification. In: Proceedings of the international conference on informatics and analytics, pp 1–8

21. Nair LR, Shetty SD, Shetty SD (2017) Streaming big data analysis for real-time sentiment based targeted advertising. Int J Elect Comput Eng 7(1):402

22. What is NoSQL? https://aws.amazon.com/nosql/

23. Vashishth V, Chhabra A, Sharma DK (2019) GMMR: a Gaussian mixture model based unsupervised machine learning approach for optimal routing in opportunistic IoT networks. Comput Commun, Elsevier 134(15):138–148. https://doi.org/10.1016/j.comcom.2018.12.001

24. Shvachko K, Kuang H, Radia S, Chansler R (2010) The hadoop distributed file system. In: 2010 IEEE 26th symposium on mass storage systems and technologies (MSST). IEEE, pp 1–10

25. Apache Hadoop. https://hadoop.apache.org/

26. Vavilapalli VK, Murthy AC, Douglas C, Agarwal S, Konar M, Evans R, Graves T et al (2013) Apache Hadoop yarn: yet another resource negotiator. In: Proceedings of the 4th annual symposium on cloud computing, pp 1–16

27. Condie T, Conway N, Alvaro P, Hellerstein JM, Elmeleegy K, Sears R (2010) MapReduce online. NSDI 10(4):20

28. Apache Hadoop Ozone. https://hadoop.apache.org/ozone/

29. HDFS Architecture Guide. https://hadoop.apache.org/docs/r1.2.1/hdfs_design.html

30. Chhabra A, Vashishth V, Sharma DK (2017) A fuzzy logic and game theory based adaptive approach for securing opportunistic networks against black hole attacks. Int J Commun Syst, Wiley. https://doi.org/10.1002/dac.3487

31. MapReduce Tutorial. https://hadoop.apache.org/docs/r1.2.1/mapred_tutorial.html

32. Mell PM, Grance T (2011) SP 800-145. The NIST definition of cloud computing. Technical report. National Institute of Standards & Technology, Gaithersburg

33. Gupta R, Gupta H, Mohania M (2012) Cloud computing and big data analytics: what is new from databases perspective? In: Srinivasa S, Bhatnagar V (eds) Big data analytics. BDA 2012. Lecture notes in computer science, vol 7678. Springer, Berlin, Heidelberg

34. Sharma DK, Pant S, Sharma M, Brahmachari S (2020) Cryptocurrency mechanisms for blockchains: models, characteristics, challenges, and applications. In: Handbook of research on blockchain technology. Academic Press, Elsevier, pp 323–348

35. Xu J, Huang E, Chen C-H, Lee LH (2015) Simulation optimization: a review and exploration in the new era of cloud computing and big data. Asia Pac J Opera Res 32:1550019. https://doi.org/10.1142/S0217595915500190

36. Zhang L-J (2012) Editorial: big services era: global trends of cloud computing and big data. IEEE Trans Serv Comput 5:467–468. https://doi.org/10.1109/TSC.2012.36

37. Chhabra A, Vashishth V, Sharma DK (2017) A game theory based secure model against Black hole attacks in Opportunistic Networks. In: Proceedings of 51st annual conference on information sciences and Systems (CISS), 22–24 March 2017, Baltimore, MD, USA, pp 1–6

38. Barona R, Mary EA, Anita. (2017) A survey on data breach challenges in cloud computing security: issues and threats. In: 2017 international conference on circuit, power and computing technologies (ICCPCT). IEEE, pp 1–8
39. Singh A, Sinha U, Sharma DK (2020) Cloud-based IoT architecture in green buildings. In: Green building management and smart automation. IGI Global, pp 164–183
40. Bhardwaj KK, Khanna A, Sharma DK, Chhabra A (2019) Designing energy-efficient IoT-based intelligent transport system: need, architecture, characteristics, challenges, and applications. In: Mittal M, Tanwar S, Agarwal B, Goyal L (eds) Energy conservation for IoT devices. Studies in systems, decision and control, vol 206. Springer, Singapore, pp 209–233
41. Sharma DK, Kaushik AK, Goel A, Bhargava S (2020) Internet of things and blockchain: integration, need, challenges, applications, and future scope. In: Handbook of research on blockchain technology. Academic Press, Elsevier, pp 271–294
42. Khanna A, Arora S, Chhabra A, Bhardwaj KK, Sharma DK (2019) IoT architecture for preventive energy conservation of smart buildings. In: Mittal M, Tanwar S, Agarwal B, Goyal L (eds) Energy conservation for IoT devices. Studies in systems, decision and control, vol 206. Springer, Singapore, pp 179–208
43. Ahmed I (2019) A brief review: security issues in cloud computing and their solutions. Telkomnika 17(6):2812–2817
44. Gatchin Y, Kolesnikova S, Kulikov R, Kainov N (2016) Insider threat towards cloud computing. In: Приоритеты мировой науки: эксперимент и научная дискуссия, pp 15–20
45. Rengaraju P, Raja Ramanan V, Lung C-H (2017) Detection and prevention of DoS attacks in software-defined cloud networks. In: 2017 IEEE conference on dependable and secure computing. IEEE, pp 217–223
46. Shaikh AA (2016) Attacks on cloud computing and its countermeasures. In: 2016 international conference on signal processing, communication, power and embedded system (SCOPES). IEEE, pp 748–752
47. Jabir RM, Khanji SIR, Ahmad LA, Alfandi O, Said H (2016) Analysis of cloud computing attacks and countermeasures. In: 2016 18th international conference on advanced communication technology (ICACT). IEEE, pp 117–123
48. Chou T-S (2013) Security threats on cloud computing vulnerabilities. Int J Comput Sci Inf Technol (IJCSIT) 5(3)
49. Acharjya DP, Ahmed K (2016) A survey on big data analytics: challenges, open research issues and tools. Int J Adv Comput Sci Appl 7(2):511–518
50. Hariri RH, Fredericks EM, Bowers KM (2019) Uncertainty in big data analytics: survey, opportunities, and challenges. J Big Data 6:44. https://doi.org/10.1186/s40537-019-0206-3
51. Dillon T, Chen W, Chang E (2010) Cloud computing: issues and challenges. In: 2010 24th IEEE international conference on advanced information networking and applications. IEEE, pp 27–33
52. Zhang Q, Cheng L, Boutaba R (2010) Cloud computing: state-of-the-art and research challenges. J Internet Serv Appl 1(1):7–18
53. Choo K-KR (2010) Cloud computing: challenges and future directions. Trends Issues Crime Crim Justice 400:1
54. Wei Y, Brian Blake M (2010) Service-oriented computing and cloud computing: challenges and opportunities. IEEE Internet Comput 14(6):72–75
55. Vashishth V, Chhabra A, Sharma DK (2019) A machine learning approach using classifier cascades for optimal routing in opportunistic Internet of Things networks. In: 16th IEEE international conference on sensing, communication, and networking (SECON), 10–13 June 2019, Boston, MA, USA
56. Sharma DK, Agarwal S, Pasrija S, Kumar S (2020) ETSP: enhanced trust-based security protocol to handle blackhole attacks in opportunistic networks. In: Jain V, Chaudhary G, Taplamacıoğlu M, Agarwal M (eds) Advances in data sciences, security and applications. Lecture notes in electrical engineering, vol 612. Springer, Singapore

Federated Learning Enabled Edge Computing Security for Internet of Medical Things: Concepts, Challenges and Open Issues

Gautam Srivastava, Dasaradharami Reddy K., Supriya Y., Gokul Yenduri, Pawan Hegde, Thippa Reddy Gadekallu, Praveen Kumar Reddy Maddikunta, and Sweta Bhattacharya

1 Introduction

The Internet of Things (IoT) is a rapidly increasing technology that allows electronic devices and sensors to communicate with one another via the Internet to ease our day-to-day activities. Using smart devices and the Internet, IoT furnishes innovative solutions to several challenges and issues related to numerous problems across the world [1, 2]. Recent advancements in IoT technology have led to the innovations of diversified applications in multiple sectors like healthcare, smart city, retail, process

---

The original version of the chapter has been revised. A correction to this chapter can be found at https://doi.org/10.1007/978-3-031-28150-1_13

G. Srivastava
Department of Mathematics and Computer Science, Brandon University, Brandon, MB, Canada
e-mail: srivastavag@brandonu.ca

D. R. K. · S. Y. · G. Yenduri · P. Hegde · P. K. R. Maddikunta · S. Bhattacharya
School of Information Technology and Engineering, Vellore Institute of Technology, Vellore, Tamil Nadu, India

Department of Mathematics and Computer Science, Brandon University, Brandon, MB, Canada
e-mail: dasaradharami.k@vit.ac.in; supriya.d@vit.ac.in; gokul.yenduri@vit.ac.in; pawan.hegde@vit.ac.in; praveenkumarreddy@vit.ac.in; sweta.b@vit.ac.in

T. R. Gadekallu (✉)
School of Information Technology and Engineering, Vellore Institute of Technology, Vellore, Tamil Nadu, India

Department of Electrical and Computer Engineering, Lebanese American University, Byblos, Beirut, Lebanon
e-mail: thippareddy.g@vit.ac.in

G. Srivastava et al. (eds.), *Security and Risk Analysis for Intelligent Edge Computing*, Advances in Information Security 103, https://doi.org/10.1007/978-3-031-28150-1_3

67

Table 1 Summary of important works on federated learning and edge computing

Ref No	Key Topic	Privacy Preserving	Reduce Latency	Data Sharing	Effective Communication	Remarks
[28]	FL and Edge	MEDIUM	LOW	HIGH	LOW	-Proposed a flexible global aggregation procedure, which adjusts the global aggregation frequency to ensure systematic use of available resources
[29]	FL and Mobile Edge	LOW	LOW	LOW	HIGH	- Proposed integrated framework with Deep Reinforcement Learning and FL to optimize the caching and offloading in User Equipment's
[30]	FL and Mobile Edge	LOW	MEDIUM	LOW	HIGH	- Proposed edge federated learning (EdgeFed) to separates the updating process of local models from mobile devices.
[31]	FL,Blockchain and Edge	HIGH	HIGH	MEDIUM	HIGH	- Proposed an innovative FLchain architecture for edge computing networks
[32]	FL and Edge	HIGH	MEDIUM	LOW	LOW	- Proposed a FL based asynchronous privacy preserving mechanism.
[33]	FL and Edge	HIGH	MEDIUM	LOW	MEDIUM	- Proposed Privacy-aware Service Placement (PSP) procedure to solve the problem of service placement with privacy awareness.
[34]	FL and Edge	MEDIUM	HIGH	LOW	MEDIUM	- Proposed a selective model aggregation approach to select local models.

automation, logistics. IoMT is a blend of IoT communication protocols with medical devices, to impart personalized healthcare services by providing remote access to patients' health conditions such as persistent illness management, blood pressure, heartbeat [3–6]. However, IoMT devices like wearable sensors are used to collect real-time health data for intelligent data analytics in the medical field.

Machine learning (ML) has advanced drastically over the last two decades, from laboratory curiosity to a practical technology in widespread commercial use. However, recent works have shown that ML is a promising technology to identify important characteristics from a complex data set and reveal their importance. The main aim of IoMT is to elevate the patient's satisfaction and quality of care provided by the healthcare providers. The data collected by the IoMT devices may consist of medical records, images, physiological signals, and evaluation results, but the ML algorithms will assist to improve decision-making, diagnosis, treatment, speed up drug research, and facilitate doctors in their processes [7, 8]. Hence, the dynamic intelligent model for data collection, analysis, and prediction can be developed using IoMT with ML. These traditional ML techniques involve a data pipeline that will combine all the data at a central server resulting in patients' data privacy risk [9].

FL is an emerging ML technology that enables devices to learn collaboratively from a shared model [10–15]. FL incorporates the basis of focused data collection, minimization, easing the privacy risks and costs resulting from traditional, centralized ML approaches [16–22]. FL can train a model, leveraging the personal data of patients without ever sharing it with other entities. Although FL is designed for securing privacy risks of individuals, we may face unique challenges like efficient communication across the federated network, managing heterogeneous systems in the same network. Moreover, communication connectivity issues are resolved by offloading the excessive computational task from IoT devices to Edge nodes by considering federation and complex resource management in real-time. Edge computing assists in rapid processing and low latency concerns of intelligent IoT applications can be achieved [23–27] (Table 1).

1.1 Comparison and Contributions

Motivated by the current progress of FL and Edge in IoMT, few reviews of related work are presented. These three topics are usually studied separately. We initially focus on relevant work in IoMT.

IoT is a large group of devices around the world that are connected to each other through the Internet and share data. These devices share real-time data among them without human intervention. IoMT is the interconnection of medical devices, hardware systems, and software applications. It is sometimes referred to as IoT in healthcare. IoMT enables wireless communication and remote devices with the Internet and to safely transfer the data to the server. IoMT majorly focuses on medical and healthcare applications like shown in Fig. 1. IoMT being able to handle sensitive data should be secured more than the IoT devices. For example, Giri et al. [35] studies about the increasing impact of IoMT in the Indian health care field and also try to improve the efficiency in managing healthcare. Some statistical tools like multiple regression and exploratory factor analysis are implemented in this study. Meanwhile, Joyia et al. [36] discusses the contribution of IoT in the healthcare

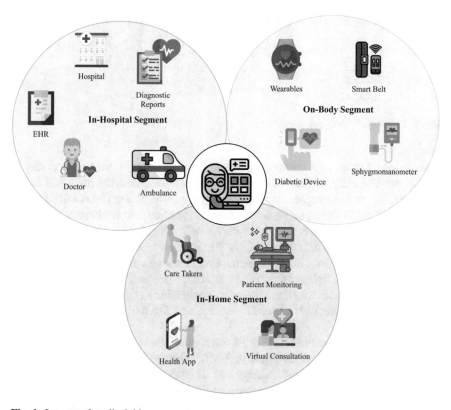

Fig. 1 Internet of medical things

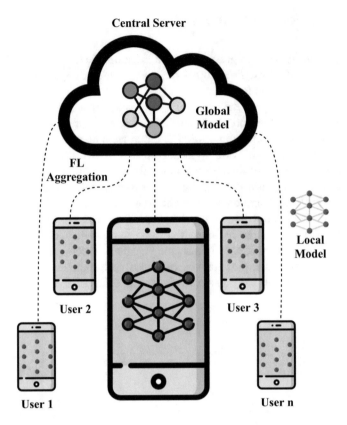

Fig. 2 Federated learning architecture

domain, application, and future challenges of IoT in terms of medical services in healthcare. Similarly, Gatouillat et al. [37] focused on improving the IoMT using some formal methodologies, and also the application of the democratization of medical devices is done practically for both the patients and healthcare providers. Aman et al. [38] discusses that when new IoMT technologies are combined with Artificial Intelligence (AI), Big Data, and Blockchain, more viable options emerge. The architecture of IoMT, applications and developments in the security area concerning IoMT in fighting COVID-19 are also discussed. This paper also gives insights into particular IoMT architecture, applications of IoMT that are coming into the light, measurements of IoMT security, and also the path for IoMT systems to combat COVID-19.

Several surveys have been done in FL during the previous few years such as [39–42]. FL is a distributed privacy-preserving ML technique. Unlike the traditional ML models, FL decentralizes the training data. It enables the devices to collectively learn from the shared model while keeping the data on the device itself. Figure 2 shows the architecture of FL. Bonawitz et al. [39] propose an extensive system

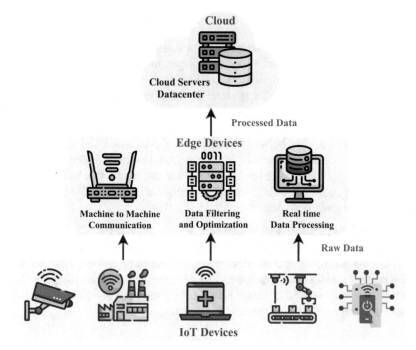

Fig. 3 Edge computing

for FL based on tensor flow in the mobile device field. The final high-end design, challenges, and relevant solutions, and future directions are also discussed. Yang et al. [40] introduces an exhaustive secure FL system, different types of FL like horizontal, vertical, and federated transfer learning. Also, a business construct based on a federated process is proposed which is an effective way to share knowledge without risking user privacy. In addition, data networks that are built on federated mechanisms in organizations are also discussed in the specified paper. Konečný et al. [41] focuses on how to reduce the communication costs by using a smaller number of variables, the model updates are quantized, arbitrary rotations, sampling before server communication. The method proposed in the specified study tries to reduce communication costs to some extent. Kourtellis et al. [42] discusses Federated Learning as a service (FLaaS) and a system that enables the collaboration of FL with third-party applications. This study explains the implementation of FLaaS in multiple environments. The privacy and permission management in FLaaS is handled in this paper.

Further, we focus on work related to edge computing. Edge computing is a distributed computing framework that brings enterprise applications closer to the data centers. These data centers can be IoT devices or local edge servers like in Fig. 3. This proximity to the data source can bring benefits in business like faster understanding, better response time, and improved bandwidth availability.

Cao et al. [43] discusses the advantages of edge computing over cloud computing, the architecture of edge computing, and applications of edge computing. Khan et al. [44] surveyed recent advances in edge computing, the importance of edge computing in real-life scenarios, and challenges in edge computing. This paper elaborates few use cases of edge computing like cloud offloading, smart home, smart city, and collaborative edge. Edge computing has enormous benefits in real-life scenarios like in manufacturing, farming, network optimization, workplace safety, improved healthcare, transportation, retail. Also, edge computing solves some cloud computing limitations like bandwidth, latency, and congestion problems.

Unique from the previous works, we aim to frame a survey on the use of FL and edge in IoMT applications. To the best of my understanding, this would be a fresh attempt on applying FL and edge computing together in IoMT. The following is a summary of the main contributions made by our work:

– We discuss the fundamentals of edge computing, FL and IoMT.
– We review the motivations and requirements of using FL in edge, applications of FL, and edge computing in healthcare and services of FL and edge duo for IoMT.
– We discuss a few challenges associated with present studies related to FL and edge computing in IoMT, further also focus on the open research problems that pave a way for researchers working in this path.

1.2   Motivations for Using FL in Edge Computing

Edge computing is an approach that is extended from cloud computing which leverages the same concept but has its advantage like mitigating latency, resource usage, energy usage, and so on. FL is just an algorithm or a kind of approach which empowers edge computing by applying the technique of model iteration instead of fetching data from the device. It also expels privacy concerns in edge computing. FL and edge computing are individually potential enough to solve many problems, but together they can create wonders in the world of technology. Figure 4 elaborates the motivation behind integrating FL and edge computing. The amalgamation of FL and edge computing provides insights to solve a few problems like:

A. Communication Problem Mitigation
One of the common challenges encountered in FL is the fact that the requirement for huge amounts of data to be communicated between users and the central server across the often expensive up-link route. In FL models, training task is typically time-consuming and resource hungry. It also leads to privacy and communication latency problems since large amounts of data are transmitted to the central server. This study proposes a new framework called federated edge learning which aggregates the local updates at the network edge [45]. A new criterion for estimating the performance called learning efficiency is defined. Every user device is facilitated with a Central Processing Unit (CPU) where the training optimization problem is formulated. The specified algorithm can minimize training time and improve learning accuracy. This algorithm is applicable in more general systems.

Fig. 4 Motivations and requirements of using FL in edge

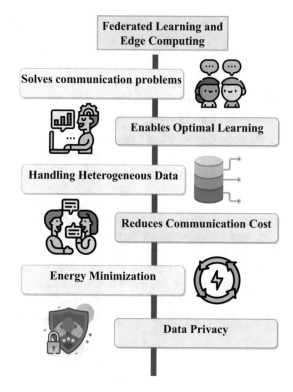

B. Optimal Learning

ML is a general term that incorporates everything from simple data summarization to multi-class categorization using support vector machines (SVMs) and deep neural networks (DNN). FL techniques are frequently used to evaluate vast volumes of data and obtain useful information for detection, categorization, and prediction of future events. Wang et al. [28] addresses the problem of how to utilize the limited communication and computation resources efficiently through a control algorithm, which is a great commutation between the local update and the global parameter optimization. The problem of how to make the most of the limited processing and communication resources available at the edge to achieve the best learning results is solved. The typical edge computing architecture is considered which is connected with the remote cloud through multiple network elements like gateways ad routers. The results of the experiments reveal that the proposed method performs close to the best with a variety of ML models and data distributions.

C. Heterogeneous Data Handling

Massive amounts of data collected by mobile clients can help to promote ML technologies. Various users have varied data sizes, resulting in an uneven scale of data controlled by edge clients. Appropriate edge client selection can significantly increase the global model's convergence rate. Mobile edge clients, on the other hand, have a finite amount of energy. To mitigate the influence of data heterogeneity, a cluster-based clients selection approach is proposed that can build a federated virtual data set that fulfills the global distribution, and also the suggested scheme can converge to an approximation optimal solution [46]. An auction-based clients selection scheme is discussed inside each cluster based on the clustering approach that fully incorporates the system's energy heterogeneity and delivers the proposed scheme's Nash equilibrium solution for balancing energy consumption and enhancing the convergence rate.

D. Communication Cost Reduction

FL is a widely used framework that protects the privacy of the client without exchanging raw data. FL faces some serious drawbacks like the high cost of communication especially when there is limited bandwidth. FedAvg algorithm has been a fascinating subject of study but still, it has some lingering issues like it takes more rounds when working with non-independent, identically distributed client data, and also the communication cost per round is very high. Mills et al. [47] proposes a novel method called Communication-Efficient FedAvg (CE-FedAvg) which subsides the number of rounds of convergence and the communications cost. Figure 5 is the Edge Federated framework. Yang [48] proposes a novel framework called RingFed which reduces the communication overhead in the training process. Rather than transferring the parameters between the client and the server the

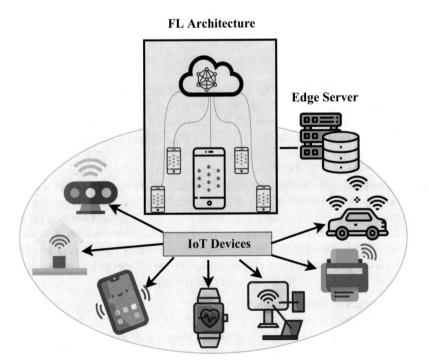

Fig. 5 Federated learning in mobile edge computing

RingFed transfers the updated parameters between the client and server. Instead of sending the updates to the server and aggregating it to the central server the aggregation of the updates is done at the proxy client. Though FL in mobile edge computing is upsurging there are some communication challenges to be dealt with. The existing FL systems which are communication efficient optimize the learning process but are not worried about the hurdles on the network. This paper proposes a Device-to-Device (D2D) communication that suits mobile edge networks [49].

E. Energy Minimization

FL includes the exchange of a learning model between users and a central server, with the learning accuracy level dictating both computation and communication latencies. Meanwhile, because wireless users have a restricted-energy budget, both local computation and transmission energy must be addressed during the FL process. The goal of this joint learning and communication challenge is to reduce the total energy consumption of the system while keeping a latency limitation in mind. Mo et al. [50] sketches a federated edge learning system where the data samples are trained on the set of edge devices that are coordinated by the edge server. A computation and communication design are explored to improve the system's energy efficiency. Two transmission protocols namely non-orthogonal multiple access (NOMA) and time division multiple access (TDMA) are regarded to transfer the ML parameters from edge devices to edge servers. The energy optimization techniques are then compared with each other to let the best be known and to be implemented.

F. Data Privacy

Privacy-preserving is important in designing, implementing, and interpreting health studies. Smart city sensing applications handle a large amount of data from the sensors. In this case, privacy is not ensured completely as the data is constantly exposed to the network. Putra et al. [51] has proposed a federated compressed learning framework, which couples the data generation along with privacy. This scheme follows the very basic ideas of compression techniques, regional joint learning along secure data exchange. The data consumption is reduced and data security is increased through this approach.

FL is an enabling technology that supports strict data privacy legislation and growing privacy concerns. FL is an adding plug to the mobile edge networks. However, large scale and complex mobile edge networks, heterogeneous devices lead to several challenges in communication costs, resource allocation, privacy, and security in FL implementation [52]. This paper presents the applications of FL for mobile edge network optimization.

Mobile edge computing (MEC) is visualized as an upsurging paradigm to handle large volumes of data that is generated from enormous mobile devices to enable intelligent services along with AI. Data privacy and high communication overheads are making MEC a weak point. In this scenario, FL has been suggested to enable collaborative data training without revealing their data, which is considered a privacy enhancement. To further improve the security and scalability of FL implementation, blockchain is considered to be effective and this leads to a new paradigm called FLChain [31]. This study mainly exposes the design of FLchain, key solutions for the design, and also the research challenges and future directions in FLchain.

Edge computing requires the individual mobile users to upload the raw data to a central server for processing. The data may be sensitive which users don't need to reveal. Liu et al. [53] attempt to keep the data of the edge devices and end-users on local storage for enabling user privacy. FL and edge computing are integrated to propose a P2FEC framework. This study also tells that the proposed framework works well when compared to standard edge computing.

1.3   Applications of FL and Edge

IoT is penetrating every aspect of present-day lives and accordingly, many intelligent applications are being created. FL can handle the data effectively generated by several IoT devices and prevent data leakage. However, the complex IoT environment creates great hurdles to the conventional FL model and paves the way to the necessity of cloud-edge architecture for intelligent IoT applications. To survive the heterogeneity issues in IoT environments, we explore the personalized FL framework. The demand for fast processing and low latency intelligent IoT applications can also be achieved with the help of edge computing [54]. FL and edge services can be utilized in a variety of situations where privacy and resource efficiency are important. We shall go through a few edge-federated learning scenarios and some recent work that's been done [32].

Vehicular edge computing (VEC) targets utilizing the communication and computation resources at the vehicular network edge. The rising demand for AI applications in intelligent connected vehicles is satisfied by FL in VEC. Image classification is one such AI application in VEC where the differences and computation efficiency in image quality influence the efficiency and accuracy of FL. Integrated local DNN models are chosen and sent to the main server by estimating the image quality and computation capability. The central server is not aware of the image and computation in vehicular clients, where privacy is protected using the FL framework. The information asymmetry is solved by a greedy algorithm [34].

The rapid increase of sensors and IoT devices along with AI gave a path for the so-called smart environments. These environments suffer from high latency and high data transmission time. da Silva et al. [55] sketches an FL architecture for traffic assessment in real-time which is assisted by Roadside Units (RSU's). This ensures that the learning is done on the edge server and also has low latency and less bandwidth usage. This study also investigates the requirements and tools for FL implementation in estimating the real-time traffic and also how this solution can be evaluated using VANET and network simulators.

The data collected in large-scale smart city sensing applications can be scarce in PM2.5 air quality monitoring systems. So, many factors affect the prediction system like poor communication, shattered records, and data privacy [51]. The novel edge computing framework namely Federated compressed learning (FCL) enables effective data generation and also assures data privacy. This study improves a green-energy-based wireless sensing network system using the FCL edge computing framework. The proposed framework is tested and proved to be effective.

2 FL and Edge Computing for Healthcare

FL has emerged as a promising method for developing cost-effective smart healthcare systems that are also more private. FL is a technique that allows high-quality AI models to be trained by averaging local updates from several health data clients, such as IoMT devices, without requiring direct access to the local data. This may restrict the disclosure of sensitive user information and preferences, reducing the danger of privacy leakage. Furthermore, because FL uses large computation and dataset resources from a variety of health data clients to train AI models, the quality of the health data training, such as accuracy, would be significantly improved, which would not be possible with centralized AI approaches that use fewer data and have limited computational capabilities.

Edge computing devices are now being utilized to constantly monitor patients, automate the delivery of health services, employ AI to improve diagnosis speed and accuracy, follow vaccination supply chains, and much more. Every day, healthcare institutions must cope with an expanding amount of digital data. However, depending on cloud or on-premises servers limits data transmission speeds, bandwidth, privacy and security concerns, and cost. All of these issues are addressed by the capacity to process data locally at the edge.

With the advent of wearables, Edge computing, and IoT the healthcare concentration is moving towards digital health. Hakak et al. [56] proposes an edge-assisted data analytics framework. This approach could make use of pre-trained algorithms to derive user-specific insights while protecting user privacy and cloud resources. Edge computing, on the other hand, achieves low latency and greater privacy and data security by storing and processing data locally or closer to edge devices. FL is increasingly being deployed at the edge, allowing consumers and businesses to benefit from and explore new opportunities in a variety of application areas like automotive, security, surveillance, and other smart city services.

Edge federated learning tackles the data problem by fully utilizing the enormous potential of data on terminal devices while maintaining user privacy, and it dramatically increases model learning performance in edge computing systems. It can be utilized in a variety of situations where privacy and resource efficiency are important. Edge federated learning can help to overcome the problem of privacy which enables the medical institutions to communicate without sharing the data related to the deceased and also meets the requirements of data privacy protection and the Health Insurance Portability and Accountability Act (HIPAA) [57]. The need for a real-time data set cannot be sometimes demystified about the sensitivity

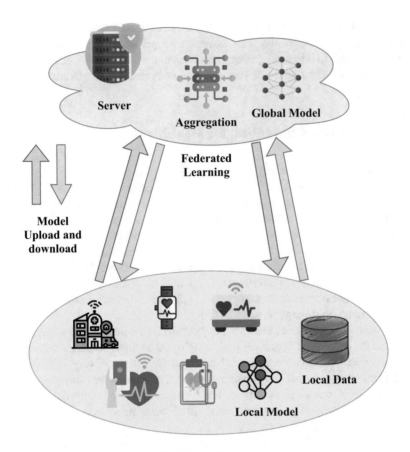

Fig. 6 Federated learning in IoMT

and privacy of medical data. Figure 6 explains the application of FL in IoMT. Also, the edge federated learning models are communication and computation efficient. Edge federated learning is a privacy-preserving ML architecture that distributes data among a large number of resource-constrained edge devices. It uses the same training technique as baseline FL, which involves an edge server distributing an initial model to each edge node, which updates the model (local model) independently using local data, with the global model updated by aggregating a subset of the local models. The following section introduces the services of FL and Edge for IoMT.

3 FL and Edge IoMT Services

We present a state-of-the-art survey on the use of FL and edge computing for IoMT services, which includes:

- FL and Edge-based techniques for privacy and security in IoMT.
- FL and Edge for reducing latency in IoMT.
- FL and Edge serving as an alternative to IoMT Data Sharing.
- FL and Edge for reducing communication cost in IoMT.
- FL and Edge for effective communication in IoMT.

3.1 FL and Edge-Based Techniques for Privacy and Security in IoMT

IoMT is a network of connected sensors, health devices, clinical systems, and wearable devices in which the advantages and security and privacy problems coexist. FL, which is an emerging technology in AI is designed to conduct ML efficiently across multiple participants or computing nodes while maintaining the confidentiality of their private information. FL can help to protect the privacy of patient data by preventing sensitive information from being exposed to prospective intruders, such as hackers. FL alone cannot solve some key parameters of smart healthcare like security, data authentication, and low latency. Edge and FL duo deployment can alleviate these difficulties. Figure 7 indicates the application of FL and edge in IoMT. FL is traditionally implemented by a parameter server and multiple edge nodes [51]. In the parameter server, gradients are collected from each node, parameters are updated according to the optimization algorithm and global parameters are maintained. The participating nodes learn from their sensitive data independently and locally. Jin et al. [58] propose a cross-cluster FL architecture based on the cross-chain approach (CFL) for addressing data sparsity and privacy leakage while maintaining high system efficiency.

3.2 FL and Edge for Reducing Latency in IoMT

The upsurge of IoMT has led to the digitization of the health care system. The immoderate demand from the patients sometimes leads to low latency, communication overload. The communication cost of the organization might accelerate due to a large number of participants. Though edge computing assists in reducing latency, there are still some lingering issues in latency. Zheng et al. [59] discusses a gradient reduction algorithm based on federated random variance. This proposal tries to lessen the iteration count between the participant and the server even when ensuring accuracy and privacy. The method is proved by linear regression and

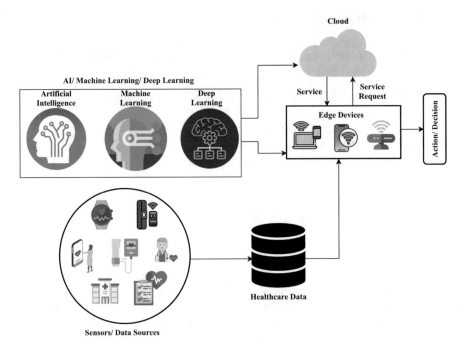

Fig. 7 FL and edge in IoMT

logistic regression at the end. The proposed process can also reduce communication expenses and make it efficient. Figure 8 indicates FL and edge for IoMT services.

3.3 FL and Edge Serving as an Alternative to IoMT Data Sharing

Many of the equipment, such as wearable gadgets, can directly communicate data with a patient's doctor, ensuring that patients receive prompt and accurate treatment. It is not often obvious who legally owns the data generated and shared by IoMT devices. The data can be used and shared in a variety of ways by other parties. The threat is amplified by the fact that the data is shared across numerous platforms, opening up multiple attack avenues. FL is getting popular as a technique to share data safely for the IoMT. A growing number of medical technologies are being combined to create a new network known as the IoMT. IoMT is projected to contribute to a valuable ML model by combining data provided by diverse devices, which can be used in a variety of scenarios, including health monitoring, auxiliary diagnosis, and pathophoresis prediction [60]. However, health-related data in an IoMT device is often sensitive to people's privacy and should not be shared or

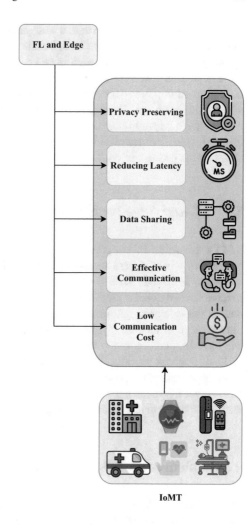

Fig. 8 FL and edge services in IoMT

aggregated casually [61]. To address this issue, FL and edge computing can be used, which allows for on-device training without transmitting data outside of the device.

3.4 FL and Edge for Reducing Communication Cost in IoMT

IoT will be used to automate healthcare facilities so that patient data may be shared and used at any time and from any location using Internet connections. Many certificateless proxy re-encryption techniques were presented to securely transfer information while reducing computation and transmission costs for data owners. Hamer et al. [62] proposed FEDBOOST and AFLBOOST, communication-efficient and theoretically-motivated ensemble algorithms for FL, where per-round communication cost is independent of the size of the ensemble. FL enables on-device training over distributed networks consisting of a massive amount of modern smart devices, such as smartphones and IoT devices. However, the leading optimization algorithm in such settings, i.e., federated averaging, suffers from heavy communication cost and inevitable performance drop, especially when the local data is distributed in a Non-IID way. Yao et al. [63] propose a feature fusion method to address this problem. By aggregating the features from both the local and global models, we achieve higher accuracy at less communication cost. Furthermore, the feature fusion modules offer better initialization for newly incoming clients and thus speed up the process of convergence.

3.5   FL and Edge for Effective Communication in IoMT

IoMT's emerging role represents a paradigm shift in every industry, particularly healthcare, where access to sensitive information from clinicians to patients and vice versa is critical. While communication is necessary to accurately transfer data to the intended destination, much of the power is dissipated in IoMT sensor nodes due to the dynamic nature of the channel. FL is upsurging with the advent of mobile Internet technology. Modern mobile gadgets have access to sensitive data and have limited computational abilities because of the restricted hardware. Ye et al. [30] proposes an EdgeFed technique that separates the process of updating the local model that is supposed to be completed independently by mobile devices. The edge server aggregates the outputs of the individual mobile devices to increase the learning speed and also decrease the global communication frequency.

4   Challenges and Open Issues

The intricacy in edge computing systems has given a boost to several technical challenges like security and privacy, scalability, heterogeneity, reliability, and resource management.

Figure 9 clusters the challenges and open issues in FL and edge computing in IoMT.

4.1   Trust on Edge Computing Systems

Any technology to be fruitful completely depends on its acceptance by the user. Trust is the major factor for a user to accept edge computing systems. Security and privacy are among the challenges faced by an edge computing system. If security and privacy issues are not addressed properly in edge computing systems then it

Fig. 9 Challenges

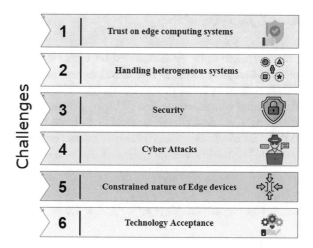

shatters the trust of a user to implement this system. The models proposed by any researcher should stimulate the consumer's trust in edge computing systems.

4.2 Handling Heterogeneous Systems

The edge computing environment consists of heterogeneous systems like edge data servers, and different cellular networks. The collaboration among multiple vendor systems is a challenging task. Interoperability, load balancing, synchronization, data privacy, resource sharing are the few factors that make collaboration in edge computing systems a challenging task.

4.3 Security

Multiple characteristics of edge computing like location-awareness, distributed architecture, mobility support, and enormous data processing will sometimes obstruct the traditional security mechanism in edge computing. To match with the mentioned security issues many edge computing mechanisms related to identity and authentication, access control systems, intrusion detection systems, privacy, trust management, visualization, and forensics, have to be developed.

4.4 Cyber Attacks

The healthcare industry is troubled by plenty of cyber security problems. These problems range from distributed denial of service (DDoS) to viruses that jeopardize the security of systems and patients' privacy. Cyber-attacks in healthcare can have far-reaching consequences that go beyond financial loss and data breach [56]. The federated edge devices are vulnerable to more attacks like Ransomware, insider threats, DDoS attacks, data breaches, business email compromise, and fraud scams, and physical attacks. Securing the devices from these attacks is much needed.

4.5 Constrained Nature of Edge Devices

A device model, it is believed, must represent knowledge of how the technology works in the real world. Edge devices pose some constraints related to storage and computational resources. As a result, unique strategies for managing and optimizing the use of local storage and computing in this environment could be investigated.

4.6 Technology Acceptance

Any device, to function properly should be able to regulate the ever-updating technology. Same way, for a federated edge device to work properly it needs to be ready to accept the updates for training purposes on regular basis. As a result, without some amount of user compliance, the model's performance will suffer significantly. As a result, it is critical that users stay involved and actively accept new technologies. It is still a struggle to persuade users to accept emerging technology for their health benefits.

5 Future Directions

5.1 Blockchain

The availability of enormous wearable devices, sensors, medical monitors led to IoMT based applications. Since there will be an enormous amount of data generated by IoMT devices, achieving privacy and security of medical data is a challenging task. Although the integration of Federated Learning and Edge computing will serve the purpose. To strengthen the privacy and security of IoMT applications we can incorporate Blockchain Technology along with Federated learning further, the system latency can be improved by incorporating Edge computing. iFLBC frameworks

[64] bring edge-AI to edge nodes by combing Blockchain and Federated Learning techniques to filter the relevant data from irrelevant data. Similarly, integration of Federated Learning and Blockchain Edge computing will lead to a new paradigm, however, FLchain [31] transforms the Mobile edge computing networks by incorporating decentralization, security, and privacy-enhancing systems.

5.2 Digital Twins

The emergence of Smart healthcare has changed the life expectance of human beings by incorporating better health operations and patients' well-being with the help of modern technologies. Using Digital Twins for personalized healthcare operations will boost the precision of identifying chronic health disorders. Combination of Federated Learning, Edge Computing with Digital twins can identify the various anomalies hence data security and privacy are improved, can reduce computation and communication costs in remote patient monitoring systems [65]. Also, a federate learning-based Digital twin edge network will fill the gap between physical systems and digital spaces ensuring an increase in data privacy and quality of service. Federated learning-based Digital Twin Wireless Networks helps in migrating the real-time data processing and computation to the edge layer, enhancing reliability, security, and data privacy.

5.3 Explainable AI

IoT has transformed the healthcare domain by introducing IoMT, however choosing analytical for distributed IoMT environment, analysis of the enormous amount of data generated by IoMT devices in a distributed environment, achieving security of IoMT devices is a challenging task. Although, recent researches on AI-enabled remote health monitoring systems were able to monitor and prevent cyberattacks. Explainable AI is promising modern technology in identifying the compromised data during cyberattacks in IoMT based patient monitoring systems by enabling caregivers to fix the problems. Also, the Federated learning-based Wearable, explainable artificial intelligence (xAI) frameworks will enable the user to have better communication using knowledge-based methods and also improve user acceptance, task performance.

5.4 Integration with 6G

The enormous usage of IoMT devices in our daily activities results in an explosive growth of data traffic, ML, and data-driven approaches. Moreover, the surge in the development of communication technology led to the innovation of 6G networks, by transforming wireless communication from "connected things" to "connected intelligence", expecting to embody advanced AI various applications promising grater-level of security and stronger privacy protections in the healthcare domain. A large amount of IoMT devices with massive data in the 6G era will force individuals to deploy efficient ML, AI-based algorithms to provide high-quality services. However implementing Federated Learning-based Edge intelligence in 6G, will bring in improved performance, ultra-low latency service, enhanced privacy of the system.

6 Conclusion

Recent advancements in IoT technology have transformed the healthcare sector, with an increasing reliance on IoMT devices to treat patients' chronic diseases. However, these IoMT devices must be more precise and faster in retrieving and sharing real-time health data from patients by maintaining privacy and security. To address the existing IoMT challenges, we discussed the possibility of integrating FL and edge computing. We started this work by providing a brief introduction to IoMT, ML, FL, and edge computing. Following that, we discussed some of the motivations for integrating FL and edge computing, as well as the applications of FL and edge computing and their integration in IoMT. To summarise, IoMT is one of the fastest-growing sectors of the IoT market, and it has the potential to transform healthcare. We also presented some challenges and issues that must be addressed, such as handling heterogeneous edge computing systems, cyber-attacks, security attacks. Also, some possible future directions have been suggested at the end of the paper.

References

1. Maddikunta PKR, Srivastava G, Gadekallu TR, Deepa N, Boopathy P (2020) Predictive model for battery life in iot networks. IET Intell Transp Syst 14(11):1388–1395
2. Maddikunta PKR, Gadekallu TR, Kaluri R, Srivastava G, Parizi RM, Khan MS (2020) Green communication in iot networks using a hybrid optimization algorithm. Comput Commun 159:97–107
3. Razdan S, Sharma S (2021) Internet of medical things (iomt): overview, emerging technologies, and case studies. IETE Technical Review, pp. 1–14
4. Swarna Priya RM, Maddikunta PKR, Parimala M, Koppu S, Gadekallu TR, Chowdhary CL , Alazab M (2020) An effective feature engineering for dnn using hybrid pca-gwo for intrusion detection in iomt architecture. Comput Commun 160:139–149
5. Xiong H, Jin C, Alazab M, Yeh KH, Wang H, Gadekallu TR, Wang W, Su C (2021) On the design of blockchain-based ECDSA with fault-tolerant batch verification protocol for blockchain-enabled IoMT. IEEE journal of biomedical and health informatics 26(5):1977–1986

6. Wang W, Chen Q, Yin Z, Srivastava G, Gadekallu TR, Alsolami F, Su C (2021) Blockchain and PUF-based lightweight authentication protocol for wireless medical sensor networks. IEEE Internet of Things Journal 9(11):8883–8891

7. Reddy GT, Reddy MPK, Lakshmanna K, Rajput DS, Kaluri R, Srivastava G (2020) Hybrid genetic algorithm and a fuzzy logic classifier for heart disease diagnosis. Evol Intell 13(2):185–196

8. Gadekallu TR, Khare N, Bhattacharya S, Singh S, Maddikunta PKR, Srivastava G (2020) Deep neural networks to predict diabetic retinopathy. J Ambient Intell Hum Comput 1–14

9. Sworna NS, Islam AM, Shatabda S, Islam S (2021) Towards development of iot-ml driven healthcare systems: a survey. J Netw Comput Appl 196:103244

10. Zhang C, Xie Y, Bai H, Yu B, Li W, Gao Y (2021) A survey on federated learning. Knowl-Based Syst 216:106775

11. Boopalan P, Ramu SP, Pham QV, Dev K, Maddikunta PKR, Gadekallu TR, Huynh-The T (2022) Fusion of federated learning and industrial Internet of Things: A survey. Computer Networks 212:109048–109058

12. Alazab M, RM, S.P., Parimala, M., Maddikunta, P.K.R., Gadekallu, T.R. and Pham, Q.V. (2021) Federated learning for cybersecurity: concepts, challenges, and future directions. IEEE Transactions on Industrial Informatics 18(5):3501–3509

13. Agrawal S, Sarkar S, Aouedi O, Yenduri G, Piamrat K, Alazab M, Bhattacharya S, Maddikunta PKR, Gadekallu TR (2022) Federated learning for intrusion detection system: Concepts, challenges and future directions. Computer Communications 195:346–361

14. Wang W, Fida MH, Lian Z, Yin Z, Pham Q-V, Gadekallu TR, Dev K, Su C (2023) Secure-enhanced federated learning for ai-empowered electric vehicle energy prediction. IEEE Consumer Electronics Magazine 12(2):27–34

15. Agrawal S, Chowdhuri A, Sarkar S, Selvanambi R, Gadekallu TR (2021) Temporal weighted averaging for asynchronous federated intrusion detection systems. Comput Intell Neurosci

16. Li Q, Wen Z, Wu Z, Hu S, Wang N, Li Y, Liu X, He B (2021) A survey on federated learning systems: vision, hype and reality for data privacy and protection. IEEE Trans Knowl Data Eng

17. Kandati DR, Gadekallu TR (2022) Genetic clustered federated learning for covid-19 detection. Electronics 11(17):2714

18. Agrawal S, Sarkar S, Alazab M, Maddikunta PKR, Gadekallu TR, Pham Q-V (2021) Genetic cfl: hyperparameter optimization in clustered federated learning. Comput Intell Neurosci

19. Javed AR, Hassan MA, Shahzad F, Ahmed W, Singh S, Baker T, Gadekallu TR (2022) Integration of blockchain technology and federated learning in vehicular (iot) networks: A comprehensive survey. Sensors 22(12):4394

20. Mothukuri V, Parizi RM, Pouriyeh S, Huang Y, Dehghantanha A, Srivastava G (2021) A survey on security and privacy of federated learning. Futur Gener Comput Syst 115:619–640

21. Mothukuri V, Khare P, Parizi RM, Pouriyeh S, Dehghantanha A, Srivastava G (2021) Federated learning-based anomaly detection for iot security attacks. IEEE Internet Things J

22. Połap D, Srivastava G, Yu K (2021) Agent architecture of an intelligent medical system based on federated learning and blockchain technology. J Inform Secur Appl 58:102748

23. Ren J, Wang H, Hou T, Zheng S, Tang C (2019) Federated learning-based computation offloading optimization in edge computing-supported internet of things. IEEE Access 7:69194–69201

24. Gadekallu TR, Pham Q-V, Nguyen DC, Maddikunta PKR, Deepa N, Prabadevi B, Pathirana PN, Zhao J, Hwang W-J (2021) Blockchain for edge of things: applications, opportunities, and challenges. IEEE Internet Things J

25. Prabadevi B, Deepa N, Pham Q-V, Nguyen DC, Reddy T, Pathirana PN, Dobre O et al (2021) Toward blockchain for edge-of-things: a new paradigm, opportunities, and future directions. IEEE Internet Things Mag

26. Razaque A, Aloqaily M, Almiani M, Jararweh Y, Srivastava G (2021) Efficient and reliable forensics using intelligent edge computing. Futur Gener Comput Syst 118:230–239

27. Gheisari M, Pham Q-V, Alazab M, Zhang X, Fernandez-Campusano C, Srivastava G (2019) Eca: an edge computing architecture for privacy-preserving in iot-based smart city. IEEE Access 7:155779–155786

28. Wang S, Tuor T, Salonidis T, Leung KK, Makaya C, He T, Chan K (2019) Adaptive federated learning in resource constrained edge computing systems. IEEE J Sel Areas Commun 37(6):1205–1221

29. Wang X, Han Y, Wang C, Zhao Q, Chen X, Chen M (2019) In-edge ai: intelligentizing mobile edge computing, caching and communication by federated learning. IEEE Netw 33(5):156–165

30. Ye Y, Li S, Liu F, Tang Y, Hu W (2020) Edgefed: optimized federated learning based on edge computing. IEEE Access 8:209191–209198

31. Nguyen DC, Ding M, Pham Q-V, Pathirana PN, Le LB, Seneviratne A, Li J, Niyato D, Poor HV (2021) Federated learning meets blockchain in edge computing: opportunities and challenges. IEEE Internet Things J

32. Lu X, Liao Y, Lio P, Hui P, Privacy-preserving asynchronous federated learning mechanism for edge network computing. IEEE Access 8:48970–48981

33. Qian Y, Hu L, Chen J, Guan X, Hassan MM, Alelaiwi A (2019) Privacy-aware service placement for mobile edge computing via federated learning. Inform Sci 505:562–570

34. Ye D, Yu R, Pan M, Han Z (2020) Federated learning in vehicular edge computing: a selective model aggregation approach. IEEE Access 8:23920–23935

35. Giri A, Chatterjee S, Paul P, Chakraborty S, Biswas S (2019) Impact of smart applications of IoMT (internet of medical things) on health-care domain in India. International Journal of Recent Technology and Engineering (IJRTE) 8(4):1–15

36. Joyia GJ, Liaqat RM, Farooq A, Rehman S (2017) Internet of medical things (iomt): applications, benefits and future challenges in healthcare domain. J. Commun. 12(4):240–247

37. Gatouillat A, Badr Y, Massot B, Sejdić E (2018) Internet of medical things: a review of recent contributions dealing with cyber-physical systems in medicine. IEEE Internet Things J 5(5):3810–3822

38. Aman AHM, Hassan WH, Sameen S, Attarbashi ZS, Alizadeh M, Latiff LA (2021) Iomt amid covid-19 pandemic: application, architecture, technology, and security. J Netw Comput Appl 174:102886

39. Bonawitz K, Eichner H, Grieskamp W, Huba D, Ingerman A, Ivanov V, Kiddon C, Konečný J, Mazzocchi S, McMahan B, Van Overveldt T (2019) Towards federated learning at scale: System design. Proceedings of machine learning and systems 1:374–388

40. Yang Q, Liu Y, Chen T, Tong Y (2019) Federated machine learning: concept and applications. ACM Trans Intell Syst Technol (TIST) 10(2):1–19

41. Yang Z, Chen M, Saad W, Hong CS, Shikh-Bahaei M (2020) Energy efficient federated learning over wireless communication networks. IEEE Transactions on Wireless Communications 20(3):1935–1949

42. Kourtellis N, Katevas K, Perino D, Flaas: federated learning as a service. In: Proceedings of the 1st workshop on distributed machine learning, pp 7–13

43. Cao K, Liu Y, Meng G, Sun Q (2020) An overview on edge computing research. IEEE Access 8:85714–85728

44. Khan WZ, Ahmed E, Hakak S, Yaqoob I, Ahmed A (2019) Edge computing: a survey. Futur Gener Comput Syst 97:219–235,

45. Ren J, Yu G, Ding G (2020) Accelerating dnn training in wireless federated edge learning systems. IEEE J Sel Areas Commun 39(1):219–232

46. Lu R, Zhang W, Wang Y, Li Q, Zhong X, Yang H, Wang D (2023) Auction-Based Cluster Federated Learning in Mobile Edge Computing Systems. IEEE Transactions on Parallel and Distributed Systems 34(4):1145–1158

47. Mills J, Hu J, Min G, Communication-efficient federated learning for wireless edge intelligence in iot. IEEE Internet Things J 7(7):5986–5994

48. Park S, Suh Y, Lee J (2021) Fedpso: federated learning using particle swarm optimization to reduce communication costs. Sensors 21(2):600

49. Zhang X, Liu Y, Liu J, Argyriou A, Han Y (2021) D2d-assisted federated learning in mobile edge computing networks. In: 2021 IEEE wireless communications and networking conference (WCNC). IEEE 2021:1–7

50. Mo X, Xu J (2021) Energy-efficient federated edge learning with joint communication and computation design. J Commun Inform Netw 6(2):110–124

51. Putra KT, Chen H-C, Ogiela MR, Chou C-L, Weng C-E, Shae Z-Y et al (2021) Federated compressed learning edge computing framework with ensuring data privacy for pm2. 5 prediction in smart city sensing applications. Sensors 21(13):4586

52. Lim WYB, Luong NC, Hoang DT, Jiao Y, Liang Y-C, Yang Q, Niyato D, Miao C (2020) Federated learning in mobile edge networks: a comprehensive survey. IEEE Commun Surv Tutorials 22(3):2031–2063

53. Liu G, Wang C, Ma X, Yang Y (2021) Keep your data locally: federated-learning-based data privacy preservation in edge computing. IEEE Netw 35(2):60–66

54. Wu Q, He K, Chen X (2020) Personalized federated learning for intelligent iot applications: a cloud-edge based framework. IEEE Open J Comput Soc 1:35–44

55. da Silva MVS, Bittencourt LF, Rivera AR (2020) Towards federated learning in edge computing for real-time traffic estimation in smart cities. In: Anais do IV Workshop de Computação Urbana. SBC 2020:166–177

56. Hakak S, Ray S, Khan WZ, Scheme E (2020) A framework for edge-assisted healthcare data analytics using federated learning. In: 2020 IEEE international conference on big data. IEEE 2020:3423–3427

57. Kaissis GA, Makowski MR, Rückert D, Braren RF (2020) Secure, privacy-preserving and federated machine learning in medical imaging. Nat Mach Intell 2(6):305–311

58. Jin H, Dai X, Xiao J, Li B, Li H, Zhang Y (2021) Cross-cluster federated learning and blockchain for internet of medical things. IEEE Internet of Things Journal 8(21):15776–15784

59. Zheng X, Shah SBH, Ren X, Li F, Nawaf L, Chakraborty C, Fayaz M (2021) Mobile Edge Computing Enabled Efficient Communication Based on Federated Learning in Internet of Medical Things. Wireless Communications and Mobile Computing 1:1–10

60. Pustokhina IV, Pustokhin DA, Gupta D, Khanna A, Shankar K, Nguyen GN (2020) An effective training scheme for deep neural network in edge computing enabled internet of medical things (iomt) systems. IEEE Access 8:107112–107123

61. Alsubaei F, Abuhussein A, Shiva S (2017) Security and privacy in the internet of medical things: taxonomy and risk assessment. In: 2017 IEEE 42nd conference on local computer networks workshops (LCN Workshops). IEEE 2017:112–120

62. Hamer J, Mohri M, Suresh AT (2020) Fedboost: a communication-efficient algorithm for federated learning. In: International conference on machine learning. Proc Mach Learn Res 2020:3973–3983

63. Yao X, Huang T, Wu C, Zhang R, Sun L (2019) Towards faster and better federated learning: a feature fusion approach. In: 2019 IEEE international conference on image processing (ICIP). IEEE 2019:175–179

64. Doku R, Rawat DB (2020) Iflbc: on the edge intelligence using federated learning blockchain network. In 2020 IEEE 6th international conference on big data security on cloud (BigDataSecurity), IEEE international conference on high performance and smart computing (HPSC) and IEEE international conference on intelligent data and security (IDS). IEEE 2020:221–226

65. Abdellatif AA, Mhaisen N, Mohamed A, Erbad A, Guizani M, Dawy Z, Nasreddine W (2022) Communication-efficient hierarchical federated learning for IoT heterogeneous systems with imbalanced data. Future Generation Computer Systems 128:406–419

Embedded Edge and Cloud Intelligence

Koushik A. Manjunatha and Sumathi Lakshmiranganatha

1 Introduction

The proliferation of artificial intelligence (AI), mobile communication technology, and high performance computing platforms are driving more advanced technologies in the field of healthcare, industrial automation, transportation, etc. The enormous data generated from various sources such as sensors, mobile users, and internet-of-things (IoT) devices not only require high bandwidth to transmit data but also require instantaneous decision-making capabilities to perform low-latency real-time tasks. According to Cisco, by 2021, the amount of data generated from the mobile users and IoT devices will reach up to 850 ZB with the total number of IoT devices crossing over 20 billion [1]. This leads to challenging data-driven decision-making due to data generated from heterogeneous devices.

Growing cloud-based applications are the ones that support edge intelligence (EI). These applications are brought about by a confluence of advancements in three global trends: (1) the proliferation of connected embedded devices such as sensors, mobile users, and IoT, (2) the growing need for intelligent sensing and cognitive decision-making capabilities to a broad set of problems from automation to control and diagnostics, and (3) the gigabit speed promises offered by the emerging 5G wireless infrastructure, supporting computational offloading. These directions will change the nature of cloud workloads, motivating new research challenges.

K. A. Manjunatha
Idaho National Laboratory, Idaho Falls, ID, USA
e-mail: koushik@crimson.ua.edu

S. Lakshmiranganatha (✉)
Los Alamos National Laboratory, Los Alamos, NM, USA
e-mail: sumathil@lanl.gov

© The Author(s), under exclusive license to Springer Nature Switzerland AG 2023
G. Srivastava et al. (eds.), *Security and Risk Analysis for Intelligent Edge Computing*, Advances in Information Security 103,
https://doi.org/10.1007/978-3-031-28150-1_4

The conventional approach of caching, processing, and decision-making at a centralized cloud server will not meet either user or business needs. For example, Google Assistant is based on cloud computing, and they have adopted a centralized location which requires continuous network connection. The centralized decision means it is always prone to security attacks and network failure affecting user privacy and authenticated decision-making. In addition, the enterprises and industrial authorities prefer to have local decision-making to avoid exposing data to the outside world.

To enhance data driven real-time intelligent decision-making and support heterogeneous applications, edge computation emerges as an extension of cloud intelligence. Edge computing offers virtualized data storage, caching, model training, model adaptation, and execution at the edge of the network. These are devices typically called edge servers which could be an IoT gateway, a router, or a cellular base station. End devices such as sensor nodes, mobile devices, and IoT devices, which request service from edge servers, are called as edge devices. The main advantage of EI can be categorized [1] as (1) *ultra low latency*: data driven computation and decision-making happen in the proximity of the end device, which significantly reduces the data transmission time to enable real-time responses to end devices, (2) *reduced computational overhead*: the edge devices as well as edge servers transmit data less frequently; also, the amount of data process at the edge servers will be less as opposed to the central server, and (3) *scalability*: the edge devices can utilize other edge servers as well as cloud servers if there are limited resources.

EI addresses the critical challenges of AI based applications and centralized decision-making with the confluence of edge computing and AI capabilities. EI refers to network of connected devices for data collection, caching, processing, analysis, and decision-making in proximity to where the data is collected [2]. Pushing intelligence to the edge enables enhanced quality and speed of data processing to protect privacy and security of the data. Compared to cloud-based intelligence that requires multiple end devices to upload the data to a central server and make decisions centrally, EI takes decisions locally reducing response time as well as saves on bandwidth resources. Moreover, the enterprises/users can also customize intelligent applications by training machine-learning (ML)/deep-learning (DL) models with locally generated data. It is predicted that the EI will be a vital component in a 6G network. By bringing intelligence and computation to edge, the latency can reduce up to 450% , saving energy consumption by 30%–40%. Typically, as a key performance indicator (KPI), higher importance is given to quality of experience (QoE), which can be determined as a combination of latency, network throughput, bandwidth, and energy consumption.

In reality, both edge and cloud intelligence should go hand in hand for a robust and resilient AI driven approach. Specifically, the cloud intelligence should be in sync with EI framework to perform scalable operations and periodic updating of AI models. Integrating the advantages of both cloud and edge provides global optimum services with minimum optimum response time for modern applications and services. For example, Firework and Cloud-Sea computing systems are the

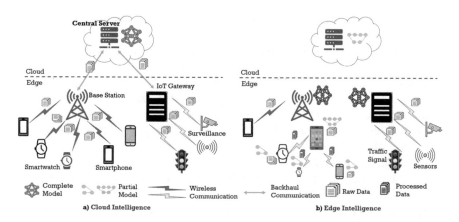

Fig. 1 Comparison of cloud intelligence (left) and EI (right)

representative systems for edge computing [3] utilizing processing power of IoT devices to filter, process, and aggregate data while employing the power and flexibility of cloud services to run complex data analytics. Amazon Web Services (AWS) can extend services seamlessly to devices so that devices can perform local operation on data they generate, while data are transferred to the cloud for processing, storage, analysis, and decision-making. A graphical representation of EI and cloud intelligence is shown in Fig. 1.

2 Cloud Intelligence

As per the National Institute of Standards (NIST) [4], "[c]loud computing is a model for enabling ubiquitous, convenient, on-demand network access to a shared pool of configurable computing resources (e.g. network servers, storage, applications, and services) that can be rapidly provisioned and released with minimal management effort or service provider interaction." The main characteristics of cloud computing include on-demand capabilities, resource pooling, rapid elasticity, and measure services. Along with data as a service (DaaS), software as a service (SaaS), platform as a service (PaaS), and infrastructure as a service (IaaS), the cloud-based intelligence service, cloud analytics as a service(CLAaaS), is enabling businesses to automate processes on an anytime-anywhere basis.

2.1 Cloud Intelligence Benefits

– **On-demand self-service:** Enterprises can expand the network, storage, power, or service on an on-demand basis without any human intervention. On-demand self-service authorizes users to request resources on run time and infrastructure transition mostly takes place immediately [3].

– **Elasticity:** Dynamic on-demand provisioning of resources improves the performance in terms of cloud computing and cloud intelligence. For cloud computing, elasticity is in relation to scaling up/down the performance (e.g., the number of processing nodes). While for CLAaaS, elasticity is the ability to dynamically bring in new data sources or models to meet emerging needs of new analyses [3].

– **Cost effective:** Pay-as-you-go models allow users to pay (a small amount) per use by effectively monitoring resource usage. This system is very transparent making the service provider and the user more comfortable to adopt it [3].

– **Resource pooling:** The computing resources from the provider are pooled to serve multiple consumers using a multi-tenant model with different physical storage and virtual machines (VMs) dynamically assigned and reassigned according to user/application demand [3].

2.2   Challenges in Cloud Intelligence

For enterprises, the data should be uploaded to cloud through its own network firewall, which raises serious concerns in terms of privacy and security. Other challenges are summarized below [5]:

– **Data quality:** The timely and accurate availability of data is crucial for decision-making. The analytical models are sensitive to data and bad data can lead to "garbage-in-garbage-out" with respect to analytical models.

– **Security and privacy:** Security is one of the major concerns with cloud intelligence. Enterprises would need to start establishing customized security protocols ensuring that data analytics are optimized and not limited because of such policies. Hacking and various attacks to cloud infrastructure would affect multiple clients even if only one site is attacked. These attacks would not only induce denial of service (DoS) but also increase latency.

– **Increased latency:** Real-time applications (ultra-low latency), which are time-sensitive, are prone to network congestion, outage, and failure. Also in a multi-tenant environment, optimal and prioritized resource allocation and load balancing will be challenging.

The are other challenges in terms of cost and bandwidth availability. These problems cannot be considered as roadblocks for every application but should be taken into account when establishing cloud intelligence.

3   Edge Intelligence

While the term EI is relatively new, the term first started in 2009 when Microsoft built an edge-based prototype to support mobile voice command recognition [2].

Currently, most organizations term EI as the paradigm of running AI algorithms locally on an end device in complement to the centralized cloud with data that are either created or collected on the device. Since EI brings data (pre-) processing and decision-making closer to the data source, delays are reduced in communication to achieve real-time requirements. EI also enables localized delay and throughput enhancements. Additionally, EI allows future applications to depend on context aware proximity and mutual detection services, device-to-device communication, and control capabilities. EI is enabled through processors, micro-controllers, storage, and connectivity modules embedded into edge devices.

3.1   Trend Drivers of Edge Intelligence

Some of the precise requirements of the current industries and enterprises driving EI include [5]:

- **Mobility:** Industries require wireless and wired networking support with high degree of mobility in terms of handovers. Managing QoE and session handover are critical aspects that can benefit from intelligence in the network components.
- **Ultra-low latency:** Decisions on detection and actuation need to be taken within a delay of less than tens of milliseconds to meet to real-time application requirements. For this, the intelligent decision-making at the edge can reduce the latency and achieve the required response time.
- **Privacy:** Industries require continuing autonomous operation without connection to core server or service to prevent damage to infrastructure. Also, industries do not prefer to expose their operation and process data to the outside world. This strongly demands EI to be at the edge of industries.
- **Security:** Security is a crucial feature that is at most a priority of every industry and enterprises to ensure access control to physical or virtual resources (e.g., data) has to be ensured. Locally provisioned and customized security protocols for authentication and authorization running on edge devices are envisioned to enable fast adaptability of the systems.
- **Prioritization:** Data transmission over the network should be adaptable and prioritized based on the latency and throughput requirement of each data transmission. The sporadic nature of data generation and transmission can efficiently be scheduled using EI.
- **Self-organization:** Transferring of operations to intelligent software from humans can be done by identifying capabilities of the devices and services and their role in infrastructure.

3.2   Scope of Edge Intelligence

EI can be a paradigm that fully exploits the available data and resources across the hierarchy of end devices, network/edge nodes, and cloud data centers to optimize the

Fig. 2 Operational hierarchy for EI

AI model training and inference. This implies an EI does not necessarily mean the AI models are fully trained at the edge but work in a cloud-edge-device coordination via data offloading. Particularly, according to the amount and path length of data offloading, the EI is categorized into six levels, as shown in Fig. 2. The description of each level of EI is given as follows [2]:

1. *Cloud intelligence:* This includes data offloading, AI model training, and inferencing fully in the cloud.
2. *Level 1– cloud-edge co-inference and Cloud training:* Data are partly offloaded to the cloud to train AI model. The trained model is shared between cloud and edge for inferencing in a cooperative manner.
3. *Level 2–in-edge co-inference and cloud training:* Data are partly offloaded to the cloud to train the AI model. The trained model is shared with the edge, and AI inferencing is done in an in-edge manner. Here, in-edge means the model inference is carried out within the network edge by fully or partially offloading the data to edge devices (via D2D communication).
4. *Level 3–on-device inference and cloud training:* Data are partly offloaded to the cloud to train AI model. Trained model is shared with the edge and inferencing is done fully local on-device manner. Here, on-devices means that no data would be offloaded.
5. *Level 4–cloud-edge co-training and inference:* AI model training and inferencing happens in the edge-cloud cooperation manner.
6. *Level 5–all in-edge:* Both AI model training and inferencing happens in the edge.
7. *Level 6–all on-device:* Both AI model training and inferencing happens in an on-device manner.

As the level of EI goes higher, the amount and path length of data offloading decreases. Consequently, the data privacy increases, and data offloading latency

Fig. 3 Architecture modes of distributed training

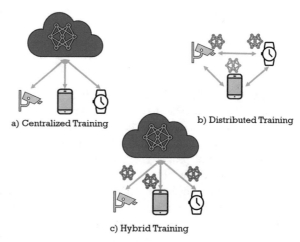

a) Centralized Training

b) Distributed Training

c) Hybrid Training

decreases with reduced bandwidth cost. In contrary, this is achieved at increased computational energy and latency. This conflict summarizes that there is no optimal level in general. The optimal level is application and industry dependent by considering latency, privacy, bandwidth, and energy efficiency.

The architecture of distributed AI training at the edge can be categorized into three modes [2] as shown in Fig. 3: centralized, decentralized, and hybrid (cloud-edge device). The cloud corresponds to central data-center, while end-devices refers to IoT devices such as cell phones, cars, sensors, etc. *(1) Centralized:* The data are gathered from the end devices then AI model training is done at the cloud. This can be identified with Levels 1, 2, and 3. *(2) Decentralized:* Each end device trains its own AI model with its local data, preserving private information locally. The global model can be obtained by establishing communication among end devices without the intervention of central data center. Level 5 corresponds to decentralized training and *(3) Hybrid (Cloud-Edge):* Edge servers (e.g., base station servers) act as hub where data from multiple end devices are gathered. Then edge servers may train the AI model by either decentralized updates or centralized training with the cloud. Levels 4 and 5 indicate the hybrid architecture.

4 Edge Training

The standard learning approach requires centralized data collection on one machine, whilst edge training relies on distributed training data on end devices and edge servers. The key idea behind edge training is to train an AI model where the data is generated or collected without necessarily sending data out of enterprise firewall to the central server. This effectively solves data privacy issues and saves network bandwidth and associated costs.

4.1   Key Performance Indicators

To better assess the performance of distributed AI model training, the following are used as KPIs [2]:

1. *Training loss:* Essentially, AI models capture the pattern of training samples. Training loss indicates how well the AI model fits the training data by minimizing the error between the predicted value and the true value. Training loss is affected by the training samples and the training methods. If the training samples and methods are not properly evaluated, then it leads to the so called *garbage-in-garbage-out* AI models.
2. *Convergence:* The distributed training of the AI models works only if a consensus on a training result is achieved by all the end-devices. The convergence defines how fast a decentralized method converges to such consensus. Under decentralized training, the convergence depends on the way error gradients were synchronized and updated.
3. *Privacy:* The raw or intermediate data should be offloaded out of the end devices to an AI model training servers or cloud. In such a scenario, it is inevitable to deal with privacy. To preserve privacy, it is expected to transfer or share less privacy-sensitive data out of end devices.
4. *Communication cost:* Training data-intensive AI models is expensive since data should be offloaded across the nodes with increased bandwidth, energy consumption, and latency. Communication cost is affected by the size of the data offloading, mode of data offloading, and available bandwidth.
5. *Latency:* Latency is arguably the most fundamental and crucial performance indicator of the EI since it indicates when the updated AI model is ready to use as well as for real-time inference. The overall latency consists of computational latency and communication latency. The computational latency depends on the capability of the nodes as well as the complexity of the AI model. The communication latency depends on the available bandwidth and size of the data being offloaded.
6. *Energy efficiency:* Most of the end devices are energy constrained hence it is desirable that AI model training is energy efficient. The AI model training at the edge devices is mainly affected by the dimension of the training samples as well as the AI model complexities.

4.2   Training Acceleration and Optimization

Training an AI model, particularly DL algorithms, requires computationally efficient resources and huge data. Often edge devices have limited computing capabilities hence result in low training efficiency. In neural network based models, the size of the network is an important factor that affects training time. Transfer learning has been the popular technique to speed up training by leveraging learned features on previous models in the current models. In an EI paradigm, features from trained model are transferred to local models, which would be re-trained with the local data. This also enhances memory and computational efficiency. In certain scenarios, interactive machine learning (iML) could accelerate the training by engaging users in generating classifiers. This approach is widely used for user preference based learning system for recommendation services. Besides, in a collaborative training environment, edge devices are enabled to communicate and learn from each other to enhance efficiency. Some of the training optimization mechanisms are discussed as follows [2, 6]:

1. *Federated learning:* Federated learning is a privacy preserving distributed AI model training and updating mechanism when data is originated by multiple end devices. Federated learning enables time and space independent and non-centralized learning. This architecture was first proposed by Google, which allows smart phones to collaboratively learn a shared model without revealing their data to each other. Then a central entity or a server combines all the model updates to a master model, which will be shared back to all the edge devices. Sharing local models and receiving master models at an end device is challenging due to an unpredictable network which poses communication breakages. Structured update and sketched update are used to update models with reduced communication cost. Structured update enables learning a model update from a restricted space parameterized using a smaller number of variables. Sketched update compresses a full model update using quantization, random rotations, and subsampling before sharing with a central server. Since a centralized server is needed to aggregate all the models, it poses a security issue. Model update without a central server can be achieved using Bayesian approach in which each device updates its belief considering its one-hop neighbor nodes. Furthermore, using blockchain technique, a secure and decentralized model update can be realized.

2. *Aggregation frequency control:* In training AI models, a common approach is to train models locally and aggregate updates centrally. Aggregation strategy significantly introduces communication overhead. Hence, aggregation content and aggregation frequency should be controlled carefully. In this regard, approximate synchronous parallel (ASP) and federated client selection (FedCS) approaches are used.

3. *Gradient compression:* To reduce communication overhead in distributed AI model training, gradient compression is another mechanism to compress the model update (i.e., gradient information). Gradient quantization and gradient sparsification are the advocated approaches. Gradient quantization compresses gradient vectors to a finite-bit low precision value. Gradient sparsification reduces communication overhead by transmitting sub-sampled gradient vectors. In addition, deep gradient compression (DGC) with a compression range of 270–$600\times$ is also advocated for wide range of convolutional neural networks (CNNs) and recurrent neural networks (RNNs). DGC employs four methods: momentum

correction, local gradient clipping, momentum factor masking, and warm-up training to preserve accuracy during compression.

4. *Deep neural network splitting:* Deep neural network splitting is aimed at protecting privacy by transmitting partially processed data rather than transmitting raw data. DNN splitting is conducted between end devices and edge server with the DNN model split between two successive layers and deployed in two partitions on different locations. To select optimal DNN splitting point differentially, a private mechanism is adopted, which also helps to feasibly outsource model training to untrusted edge servers. To preserve privacy on sensitive information, private feature extractor algorithms are also used, which outputs primary information while discarding all other sensitive information. Three different approaches such as dimensionality reduction, noise addition, and Siamese fine-tuning are adopted to make sensitive information unpredictable. These approaches also contribute in dealing with the tremendous computation of AI models. With dense deployment of end devices, parallelization approaches are also employed in terms of data parallelism and model parallelism. However, model parallelism may lead to severe under utilization of computational resources, whereas data parallelism induces increased communication overhead. Pipeline parallelism is also widely adopted as an enhancement to model parallelism, where multiple mini-batches are injected into the system at once to ensure optimal and concurrent use of computational resources.

5. *Gossip training:* This is reducing model training in a decentralized environment, by leveraging the advantages of full asynchronization and total decentralization. Gossip SGD (GSGD) is developed to train an AI model in an asynchronization and decentralized way by managing group of independent nodes, which manage an AI model. Each independent node updates the hosted AI model locally in gradient step and then shares its gradient information with another randomly selected node in mixing update step. This repeats until a consensus on AI model convergence is reached. But on a large scale systems, the gossip learning would lead to a communication imbalance, poor convergence, and heavy communication overhead. Hence, GossipGraD, a gossip communication protocol based SGD is advocated in which nodes exchange their gradient update after every $log(p)$ steps. It also considers the rotation of communication partners for gradient diffusion and sample shuffling to prevent overfitting.

5 Edge Inference

There is a rapid growth of network size with edge devices and their resource requirements. Consequently, there is an increasing gap between required computation and available computation capacity provided by the hardware architecture. After distributed training of the AI model, it is crucial to implement the model inference at the edge to enable high quality EI service. EI is usually performed on the edge devices which bounds execution time, accuracy, energy efficiency, etc.

by technology scaling. Also, edge inference acceleration is crucial to reduce the run time of inference on edge devices and achieve real-time responses without altering the structure of neural networks [1]. The inference acceleration can be divided into two categories: hardware acceleration and software acceleration. Hardware acceleration targets parallelizing inference tasks to available hardware such as CPU, GPU, and DSP. Software acceleration targets optimizing the resource management, pipeline design, and compiler. The models and frameworks that are being used in edge inference and its acceleration are discussed here.

5.1 Model Design

Complex AI models such DL models are becoming deeper, larger, and slower with more computation power. This makes it difficult to run models on edge devices such as IoT device terminals and embedded devices due to their limited capacity and power. It is also evaluated that the performance of an AI model on edge device and edge inference costs up to two orders of magnitude in greater energy and response time than central cloud intelligence. Hence, lightweight AI models are under development to enhance model decision and inference at the edge devices. Such lightweight AI models are developed with two approaches: architecture search and human-invented architecture.

Architecture Search Designing AI models for EI is quite time consuming and needs human intelligent effort. Instead, automatic architecture search is advocated in frameworks such as NASNet, AmoebaNet, and Adanet that could achieve competitive performance in classification and recognition. However, these frameworks require significant hardware configuration. Differential architecture search (DARTS) framework is developed to significantly reduce the hardware dependency by continuous relaxation of the architecture representation with gradient descent.

Human-Invented Architecture To overcome the hardware dependency, deep-wise convolutional neural-network is proposed to develop lightweight neural network such as MobileNets, but depth-wise convolutional neural network only filters input channels. Instead, combining depth-wise convolution and 1×1 point-wise convolution could overcome this drawback. Group convolution is another approach to reduce computational cost for model designing. However, the outputs of one channel is derived from the small part of input channels. This can be overcome by enabling information exchange between channels using *channel shuffle* approach. However, depth-wise convolution and group convolution are usually based on *sparsely-connected* convolutions, which may affect inter-group information exchange and degrades model performance. Hence, merging and evolution approach is employed. In merging operation, the features of the same location among different channels are merged to generate a new feature map. Whereas, evolution operation extracts the information of location from the new feature map and combines extracted information with the original network.

5.2   Model Compression

Complex AI models such as DNN are quite powerful in achieving high accuracy predictions. However edge infrastructures are limited by storage, memory, and computing powers, which makes it challenging to run complex AI algorithms such as DNN. Also, the gap between energy efficiency of DNN and hardware capacity are also growing. Hence, model compression is aimed to lighten the model, improve energy efficiency, and enhance real-time inferencing on edge devices, without compromising on the accuracy. Following sections briefly discuss several model compression techniques [1]:

1. *Low-rank approximation:* A low-rank convolutional kernel is multiplied to replace high-dimensional kernels. The approach is based on the fact that any matrix can be represented as multiplication of multiple matrices of smaller size. This approach reduces the computational complexity on the reduced matrices. Instead of compression high-dimensional kernels, network compression is also proposed with steps rank selection, low-rank tensor decomposition, and fine-tuning. Using variational Baysian matrix factorization (VBMF) the rank of each layer is determined. Then Tucker decomposition is applied to decompose the convolutional layer matrix into three components of dimension. Model compression in CNN is challenging due to large amount of convolutional operations; however, this can be overcome by using convolutional kernel separation method.

2. *Knowledge distillation:* Knowledge distillation is based on transfer learning, in which smaller neural networks are trained with the distilled knowledge from larger model. The compact model is called student model whilst the larger model is called teacher model. First, a function learned by a high performing model is used to label pseudo data. Then the labelled data is used to train a compact but expressive model, which can perform in competence with high performing model. This approach is limited to shallow neural networks. In a different approach, first train a large and complex AI model, which is an ensemble of multiple models with small and simple student models. At the softmax output, the temperature setting will be increased at the student node to train the transfer data set received from the teacher. The shallow networks can also be used as a teacher model by properly defining *attention*, a set of spatial maps that the network focuses most on in the input. The attention framework can also be used to supervise the student model.

   In an alternative approach of designing student model, the convolutional layer of teacher model is replaced with cheaper alternatives. The newly generated student model is trained under the supervision of the teacher model. The newly generated student model learns parameters through layer-wise fine-tuning to improve prediction loss. Typically, the teacher model is trained beforehand, but the teacher and student models can be trained in parallel to reduce training latency. This can also help the student model to learn difference between output and its target, and the possible path towards the final target learnt by the teacher model.

In a privacy preserved environment, it is difficult to train the student model by sharing data. Hence, to distill learned knowledge, the metadata from the teacher model is shared to reconstruct original dataset. Then the noise associated with the data is removed through gradients, which could partially reconstruct the original data set used by the teacher model.

3. *Compact layer design:* Compact layer design in neural network is essential to reduce the resource consumption by avoiding redundant computations and weights, which end up to be close to 0. ResidualNet is an approach which replaces the fully connected layers with the global average pooling. Whereas, for a CNN model, compression is done either decomposing 3×3 convolution into 1×1 convolutions or reducing input channels in 3×3 convolutions. This reduces the quantity of parameters in the CNN model. For RNN, mining dictionary approach is used to adjust compression rate with consideration to different redundancy degrees amongst layers.

4. *Network pruning:* Network pruning deletes the redundant parameters in the neural network model. The connections with less weights or weights less than a threshold are removed, leading to a sparse network. Alternatively, the low performing neurons are deleted and using width multiplier all the layer sizes are expanded. However, the primary assumption of low weight neurons contributing less to the prediction might destroy the network structure. To overcome the limitations of thresholding, the differentiability-based pruning is adopted to optimize the weights and thresholds for each layer. Conversely, Taylor expansion is also applied by considering pruning as an optimization problem, trying to minimize weight matrix which minimizes the change in cost function. For CNN models, gate decorator is proposed which multiplies the output by channel-wise scaling factors. If the scaling factor is set to 0, then the corresponding filter would be removed. Also, energy aware pruning strategies are also proposed in which the consumed energy from hardware measurements are used to identify most energy consuming parts of CNN model and prune the weights to maximize energy efficiency. The probability based dropout approach to minimize the number of redundant hidden elements in each layer is also proposed.

5. *Parameter quantization:* It is not always necessary to have highly precise parameters in neural networks in achieving high performance, particularly when high precise parameters are redundant. For CNN, the fully connected layers always consume most of the memory. Vector quantization and hash methods along with k-means clustering are proposed to reduce parameters in CNN. Besides, network binarization is an extreme case of weight quantization, in which the weights are either mapped to two possible values (1 or -1). It is also proved that the binary weights and activations could achieve better performance than continuous weights.

5.3   Model Partition and Early Exit

To subside the model training overhead on edge devices, one intuitive solution is to partition the model by offloading the computation intensive part to the edge server by considering latency, energy, and privacy. The model partition can be categorized into two types: partition between server and device and partition between devices. For the partition between server and devices, the key is to identify the partition point to get optimal model inference performance [2]. Lossy feature encoding is proposed to transmit compressed intermediate data after model partition. Whereas, for model partition between devices, a micro-scale cluster based approach in WiFi-enabled devices for partitioned DNN model inference is used. The model which carries the DNN model is the group owner, and the others act as the slave nodes.

A DNN model with high accuracy typically has a deep structure, and it consumes more computational resource to execute such model on end device. To accelerate model inference, the model early-exit leverages output of early layers to get the model result. Specifically, the inference is finished using a partial DNN model without going through complete DNN structure to optimize latency.

5.4   Edge Caching

Edge caching is another approach to enhance the AI model inference at the edge (i.e., optimizing the latency issue by caching the AI inference results and reuse it). The model results will be stored at the cloud server, and it will be shared when the edge device requests a cached model. But this approach would still induce latency and communication overhead. Instead, the edge server can store the model results [1, 2]. Particularly, the edge servers replaces the least frequently used inferences as the model replacement strategy. As an extension, model inference can also be stored at the mobile devices with Markov chains approach to prefetch the data on to mobile devices. Considering the scenario that same the application runs on multiple devices in the close proximity and also DNN model often process similar input data, FoggyCache technique is used to reduce redundant computations.

6   Edge Offloading

Computation is a key task that enables EI. Considering the hardware limitation of the edge devices, computational offloading offers promising approaches to increase computation capability of several computationally complex applications. Designing optimal offloading strategy includes achieving objectives of energy efficiency, reduced latency, and privacy preservation. Hence, offloading can be divided into five categories as shown in Fig. 4: device-to-cloud (D2C), device-to-edge (D2E), device-to-device (D2D), hybrid architecture, and caching [1].

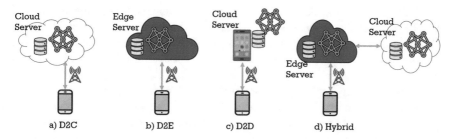

Fig. 4 Distributed caching modes

6.1 D2C Offloading

Most of computationally intensive applications such as Apple Siri and Google translation adopt pure cloud-based offloading strategy, in which the inference is done by powerful computers. But these have serious drawbacks: (1) end devices require uploading enormous data to the cloud continuously, which creates a major bottleneck due to network congestion; (2) the device need to have internet connectivity, and relative applications cannot be used when the device is offline; and (3) the uploaded data can have private information which could be vulnerable to privacy breaches. As an example, the DNN models typically have many layers which process input data layer by layer. Hence, the size of the data can be scaled down through layers. In this regards, DNN layer scheduling based offloading scheme is formulated. Edge devices lacking computational resources first upload data to near by edge server. The edge server processes original data through a few low network layers. The generated intermediate data are offloaded to the central server for further processing and generate model results. Such model partitioning and scheduling can be designed from the perspective of energy, latency as well as privacy. The easiest strategy is to leveraging inference task to the cloud when the network condition is good or else perform model compression locally. Also, it is conflicting to jointly optimize network throughput and latency on both mobile device and cloud server. In contrast, for heterogeneous sensor based applications a collective offloading scheme is considered which makes it the best possible use of various computing resources on mobile devices.

6.2 D2E Offloading

Persistent challenges of wireless network congestion and potential risk of private information leakage drives potential of D2E offloading. With D2E offloading, there are two main challenges including (1) which component of the model could be offloaded to the edge servers and (2) which edge server should be selected to offload. Model partitioning can still be used in this scenario by deploying partitioned model at edge server and edge device. As an example, the computation-intensive layers of DNN are executed on edge servers, and rest of the layers are executed on the device. To improve energy efficiency, the DNN is partitioned at the end of the convolution layers. Then output features are compressed and transmitted over to edge servers to minimize bandwidth usage. To preserve privacy, the compressed features can be encrypted using homomorphic encryption and then transmit over to the edge server. However, encryption task is also computation intensive, hence private CNN approach is used in which most inference tasks are offloaded to edge servers to avoid privacy breaches. Selection of the edge server is challenging when the edge devices are involved with mobility (for example, vehicular networking). In such scenarios, reinforcement learning-based algorithms are proposed to select optimal edge servers. In addition, a proper resource allocation and scheduling needs to be considered to establish cooperation among several edge devices that are contending for edge server.

6.3   D2D Offloading

Most of the AI models can be executed on the mobile devices after model compression. In some statics scenarios (such as smart watch), the end devices can offload data to its connected powerful devices. There are two such offloading scenarios, including binary decisions based offloading and partial offloading. Binary offloading is executing task either at local device or through offloading, while partial offloading involves dividing inference task into multiple sub-tasks and offloading them to its connected devices. Complete offloading does not necessarily outperform partial offloading since complete offloading involves transmission of complete data to the associated devices and in turn increases latency. In such cases where the connected device is not powerful enough, a cluster of edge devices could be organized as a virtualized edge server. Also, distributed solo learning enables edge devices or edge servers to train AI models using local data. Eventually, each model may become an expert at predicting local phenomenon.

6.4   Hybrid Offloading

This approach utilizes the benefits of seamless collaboration between edge and cloud computing resources by taking advantage of cloud, edge, and mobile devices in a holistic manner. All sections are jointly trained in the cloud to reduce communication overhead as well as resource usage for edge devices. During inference, each edge device performs local computation, and the results are aggregated for final output. The edge devices and edge servers have the risk of failure, which results in DNN failure. In such scenarios, the physical nodes should be skipped to minimize the impact of failed DNNs.

7   Future Directions and Open Challenges

From the past discussions, it is evident that the EI certainly benefits people with intelligent services with reduced dependency on central cloud servers with enhanced privacy. Still the EI has unresolved and open challenges [1, 2] including data scarcity at edge, data consistency on edge devices, bad adaptability of statically trained model, privacy and security issues, and incentive and business mechanism.

7.1   Data Scarcity at Edge

For an enhanced edge inference, the accuracy of the AI model is crucial. Most of AI models are supervised models and depend on high-quality training instances. However, it is often not the case that edge device will have all the training instances which is representative of all the behaviors. Also, unlike centralized intelligence, edge devices have to use self-generated data to train AI models. Most of the current solutions ignore these limitations. In addition, most of the data are unlabelled and require active learning mechanisms with human intervention, which can be possible in very few applications. Besides, federated learning is helpful for collaborative training but cannot be used for solo and personalized models. The mentioned limitations can be possibly overcome by using the following solutions:

– Adopt shallow and simple AI models which can be trained using limited data. In most of the scenarios, traditional AI models such as logistic regression, naive Bayes, and random forest perform better. It is always reasonable to start with simple AI models and increase its complexity until there is satisfactory performance in consideration to latency, energy efficiency, as well as computational resources.
– Use an incremental learning approach to re-train a pre-trained model as the new data are available. In such cases, the model can be trained with sparse and limited training instances.
– Apply data augmentation approaches to enable model to be more robust by enriching data using re-sampling approach during the training phase. Advanced DL algorithms such as generative adversarial networks (GAN) can also be used with latent sampling to generate fake data to train models.

7.2   Data Consistency on Edge Devices

EI based applications typically collect data from large amount of sensors which have different sensing environments, sensitivity levels, and operating ranges, leading to inconsistency in the data. The environment and its conditions add background noise to the sensor measurements, impacting the model's accuracy. Also due to sensor heterogeneity, the data collected from same environment can differ. This can be overcome by centrally collecting all the data and training a single model. However, this is outside the scope of EI. Alternatively, data augmentation and representation learning could be the solutions. Data augmentation could enrich the data to make model robust to noise during training. GAN networks can work in supportive of data augmentation with a latent sampling approach. Similarly, representation learning can be used to extract more effective features by hiding differences between different hardware.

7.3   Adaptability of Trained Model

Most of the ML algorithms are centrally trained or trained once. But most of the applications face data inconsistencies leading to poor performance with the trained model. In contrary, in a decentralized training only local data are used hence performance decreases as the models are used in a different environment. Therefore, continuous and incremental learning is necessary for knowledge accumulation and self-learning on new tasks. Continuous and incremental learning are not meant for edge devices, so model compression should be used every time a model is updated at the edge server or central server to be deployable at the edge devices. Knowledge sharing can be beneficial under circumstances when the request is submitted to an edge server that does not meet a required performance, then such edge servers can receive knowledge from other edge servers to meet required performance. In some cases, the task can be transferred to the edge server which can meet the required performance.

7.4   Privacy and Security Issues

To realise EI, heterogeneous edge devices, edge servers, and central cloud are required to operate collaboratively. Thus, end devices are required to share locally cached data and computing tasks with untrusted devices. The data may contain private information (e.g., tokens and images), which can be vulnerable to privacy leakage. If the data are not encrypted, the malicious users can easily extract private information. Even if the data was pre-processed, it is still possible to extract private information. In addition, the malicious users can perform a data spoiling attack and DoS attack to hamper collaborative computing. The key challenge is the lack of enhanced privacy and security preserving protocols to ensure user authentication and access control, model and data integrity, and mutual platform verification for EI.

Data encryption is one of the possible solutions, but it also requires data to be decrypted at the receiving end, leading to processing delay. To overcome processing delays, homomorphic encryption is helpful. With homomorphic encryption, direct computation can be done on encrypted data. After the decryption, the result is same as the result achieved by computation on the unencrypted data. Also, blockchain technology can be considered as one of the potential solutions. Current solutions require more computational resources with high latency. More research work is needed in this direction.

7.5 *Incentive and Business Mechanism*

Data collection, pre-processing, and AI model training are the utmost important steps for EI. In a heterogeneous deployment devices, it is always challenging to ensure quality of the data and its usability. It is not realistic to assume all the data collectors are willing to contribute unselfishly to collaborative EI model training and inference. In private deployment scenarios, all devices are inherently motivated to collaborate and develop EI model. However, with the unfamiliar participants in public scenarios, collaborative EI model development is challenging. In such scenarios all the participants need to be incentivised to collaborate and develop EI models.

As a means for collaborative EI training and inference, blockchain with a smart contract may be integrated into EI service by running on decentralized edge servers. To accommodate heterogeneous services and end devices, designing resource friendly lightweight blockchain consensus protocol for EI is highly desirable.

Reasonable incentive mechanism should be added to reward each participant's contribution and meet their requirements. Also, each operator expects to achieve high model accuracy with low cost possible. The challenges of designing optimal incentive mechanism are dependent on quantifiable contributions of each participants. Future efforts could focus on addressing some of these challenges.

8 Summary

Due to the surge of IoT based applications, there is a strong need to push AI solutions from cloud to network edge. A comprehensive discussion surrounding cloud intelligence and EI was presented. Particularly, the benefits and limitations of cloud intelligence were discussed. Then EI was discussed in detail in terms of edge training, edge inference, edge offloading, and edge offloading. Finally, the challenges and future directions associated with cloud intelligence and EI were discussed. For an enhanced performance, it is advantageous to leverage the benefits of both cloud intelligence as well as edge intelligence to operate together. Integrating the advantages of both cloud and edge provides global optimum services with minimum optimum response time for modern applications and services. Finally, hosting intelligence at cloud or edge depends on the application, business needs, as well as performance.

References

1. Xu D, Li T, Li Y, Su X, Tarkoma S, Jiang T, Crowcroft J, Hui P (2020) Edge intelligence: architectures, challenges, and applications. arXiv preprint arXiv:2003.12172
2. Zhou Z, Chen X, Li E, Zeng L, Luo K, Zhang J (2019) Edge intelligence: paving the last mile of artificial intelligence with edge computing. Proc IEEE 107(8):1738–1762
3. Balachandran BM, Prasad S (2017) Challenges and benefits of deploying big data analytics in the cloud for business intelligence. Proc Comput Sci 112:1112–1122
4. Kasem M, Hassanein EE (2014) Cloud business intelligence survey. Int J Comput Appl 90(1):23–28
5. Abdelzaher T, Hao Y, Jayarajah K, Misra A, Skarin P, Yao S, Weerakoon D, Årzén K-E (2020) Five challenges in cloud-enabled intelligence and control. ACM Trans Internet Technol (TOIT) 20(1):1–19
6. Hussain F, Hussain R, Hassan SA, Hossain E (2020) Machine learning in iot security: current solutions and future challenges. IEEE Commun Surv Tutorials 22(3):1686–1721

The Analysis on Impact of Cyber Security Threats on Smart Grids

Abhishek Vishnoi and Vikas Verma

1   Introduction

Smart grid is the next level of technological advancement in the rea of power system network. It will make new reforms in the industrial and power market with providing high grade solutions which will increase the efficiency of the conventional electric grade. The technology is the hybrid mix of digital communication technologies and the advancement computing method. This will enhance the reliability and performance of power system with the integration various renewable energy source, intelligent distribution network and balanced demand response. With keeping the view on the effective features of smart grid, the cyber security in the grid system is emerges as a sensitive and prime issue. This is because of large number of inter connected electronic devices which communicate each other using digital network [1, 2]. This communication includes all the critical information floating between the devices which will affect the reliability of complete infrastructure. Therefore, these smart networks must be heterogenic, stable and the have efficient constraints which will deploy the security over all the network segment.

Next, this chapter presents current research efforts aimed at improving the security of smart grid applications and infrastructure. Finally, current challenges are identified to facilitate future research efforts. It is often believed that existing solutions can be applied directly to new engineering domains. Unfortunately, careful study of the unique challenges posed by new disciplines reveals their specificity and often requires new approaches and solutions [3, 4]. This chapter argues that the smart grid will replace its incredibly successful and reliable predecessor and introduce a new set of security challenges that require a new

A. Vishnoi · V. Verma (✉)
Kanpur Institute of Technology, Kanpur, India

© The Author(s), under exclusive license to Springer Nature Switzerland AG 2023   111
G. Srivastava et al. (eds.), *Security and Risk Analysis for Intelligent Edge Computing*, Advances in Information Security 103,
https://doi.org/10.1007/978-3-031-28150-1_5

approach to cybersecurity. This is known as cyber-physical system security. The tight coupling between information and communication technology and physical systems raises new security concerns that require rethinking of commonly used goals and methods. Existing security approaches are either inapplicable, impractical, poorly scalable, incompatible, or simply unsuitable for addressing the challenges of very complex environments such as smart grids. Achieving the vision of a secure smart grid infrastructure requires the cooperation of the entire industry, the research community and policy makers.

2   Architecture of Communication in Smart Grid Network

The basic layout of communication network in smart grid includes various communication protocol for power grids. In electric power network system, there are much complexities such as thousands of power distribution substations, millions of customers and many distributed energy facilities. The fundamental model of smart grid includes specific domains such as bulk generation, transmission, distribution, markets, consumers, operations and service providers. The initial flow has two-way communication flow of power and information collectors for power management [5, 6]. These are interconnected through hybrid highly distributed an in hierarchical topology with a backbone network (Fig. 1).

The backbone network includes infrastructural nodes for interdomain communication. It uses bandwidth routers and local area network to pass on the information among various domains. The backbone network also uses the line technology such as optical fiber and SCADA system for high-speed communication and monitoring of operation, transmission and distribution domains. A local area network uses

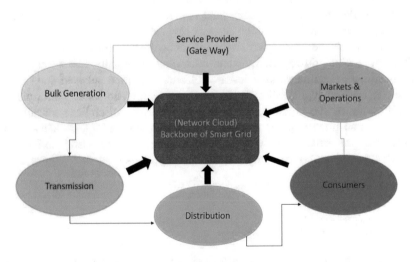

Fig. 1 Basic architecture of secure smart grid network system

wireless communication such as cellular network, sensors and intelligent devices which is also operational on the wireless network also so in comparison with conventional power system the smart grid will grip on wireless as well as line network technology for highly skilled, large scale and well operated power network infrastructure [7, 8].

3 Features of Network

In smart grid network the communication structure is much similar to the internet in the respect of hierarchy and complexity but there is certain prime difference between these two networks which are:

3.1 Performance Indicator

The basic function of internet is to provide data services with large amount of material and fairness is the prime concern but in power communication services are delivered with reliability and real time information management system in this latency is important factor.

3.2 Traffic Model

In internet the traffic flows with self similarly properly (www) traffic but in power network there is large amount of periodic traffic flow due to consistent monitoring like meter reading and data sampling is power stations as well as in home area network.

3.3 Timing Efficiency

In internet IP traffic the delay requirements are of 100–150 ms in order to provide multimedia services but in smart grid. The delay is very stringent. It is in few ns.

3.4 Communication Layout

In internet end to end principle is follows it provide peer to peer communication across the world. But in smart grid network two-way communication is follows such as top-down and bottom-up. It also supports peer to peer model also but it restricts in local area network due to security concerns.

3.5 Internet Protocol Stack

The internet is built on the IP protocol which is now at IPv6. This smart can help many protocols besides this. It will depend upon their requirement and functionalities.

4 Need and Requirements of Cyber Security in Smart Grid

To make the power system network secure and reliable operation network the exchanges of information in power infra structure must be critically secure in nature. To ensure this first we need to understand the objectives as well as requirements for the smart grid.

4.1 Needs of Cybersecurity in the Grid

There are three category of security objectives which are needed for the smart grid are as follows (Fig. 2).

4.1.1 Accessibility

In smart grid there must be ensured and reliable in the accessing and use of information because loss of information will break the chain and cause the disruption in the information which may further disrupt the power delivery.

Fig. 2 Cyber security
objectives in the power grid

4.1.2 Coherence

For the protection against manipulation of data and disruption coherence is required which will safely guard and induce the correct decision and authenticity.

4.1.3 Secrecy

To ensure privacy and to protect proprietary information there must be certain restriction which will prevent unauthorize disclosure of information among public.

4.2   Requirements of Cyber Security

The cyber security mainly deals with the security of information and network segments related with smart grid the major requirements are in the field of in operations, communication protocols networks, controllability and accessibility of devices [9, 10].

4.2.1   Operations

As the smart grid network is open to large area it has possibility for the network attacks. Therefore, network needs to be performed consistent profiling, testing and monitoring the traffic status to detect and identify the abnormal duration due to malware attacks. Resilient operation is required for the sustainable operation in the smart grid.

4.2.2   Communication Protocols

In smart grid network message delivery requires security as well as scheduled time handling in distribution and transmission network. Both the requirement must be optimal in nature.

4.2.3   Accessibility and Control

There are large number of electronic devices and users involved in the smart grid network so the identification and authentication is the main method to verify the identity of the connected devices. To protect the system from unauthorize personnel, the network must have strict control on accessing the information and the basic cryptographic functions to perform data encryption and authentication.

5   Potential Threads in the Smart Grid

To prevent the system from cyber-attacks it is essential to understand the potential vulnerable attack via communication network. The classification of attack is required to be analyzed for safeguarding the smart grid system. Cyber security attack is of two types first one is self- centric trespassers and second one is malicious users. Self-centric trespassers try to attempt more than one network resources in illegitimate ways by violating the standard communication protocols. They disrupt and manipulate the information in an illegal manner.

Malicious users attacking is the most critical issue because huge number of electronic devices and computing devices are connected for monitoring and controlling which provide the services like sharing and downloading the data. These may lead to cataclysmic damage to power supply which tends to power outages. All these possible tracks of cyber-attack are vulnerable to smart grid under denial of service (DOS) attack which will leads to degrading the performance of communication network through non operation of electronics devices.

6   Solution to the Challenges

Cyber security challenges in the smart grid network are the critical issue which require ideal attention to protect against the vulnerable threats and risks. These potential threats are major hurdles in maintain the effective power management system. There are various methods used for securing different parameters in the networks. The following possible solutions are (Fig. 3).

6.1   Security of Network

The network security issue is the main issue and commonly occurred in smart grid network. The denial of service (DOS) attack makes the system unresponsive and disrupt all the functions rapidly. This is overcome by the DOS detection and mitigations. The DOS attack is being detected through its IP packet content, pattern and properties. For this detection the method involved are using flow entropy signal strength, sensing time measurement, signatures and failure in transmission count. Once detected the smart grid network take appropriate actions with in short duration on the network layer using various techniques such as pushback, filtering, rate limiting, cleaning center and reconfiguration.

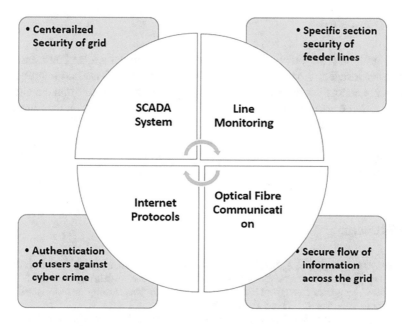

Fig. 3 Secure flow & management of power in smart grid

6.2 Security of Data

To secure the smart grid network the data and information which flows between the electronic devices are needed to be secure and the object users are also required to be authenticated. For the protection of data cryptography algorithms and methods are adopted which will encrypt the data in a secure manner for communication, authentication of user, authorization of data access to selective users. For the protection of data various methods are involved in such as encryption, remote access to VPN, authentication, time asymmetry, secret info asymmetry and hybrid asymmetry by creating channels for every transmission.

6.3 Security in Network Protocol

For any network to be protected against any cyber or malware attacks there is requirement of secure network protocols and architecture. As the existing smart grid networks use internet- based protocol for their communication infrastructure. These protocols are IP sec, TLS, DNP3. IEC61850 etc. These protocols can be made fully secure by adding one security layer to it and it will be use as end-to-end communication between different devices in the smart grid. The secure architecture can be designed as role based or trust computing-based network architecture.

7   Conclusion

Smart grids are the most advanced version of electrical grids. They are much competent efficient and productive than conventional grids. It is also integrable with renewable energy source and performs lot of functions with full transparency. As there are large number of interconnected devices and millions of data are being flowed from one to another through the internet-based protocols. There may be risk of cyber-attack and loss of data by unauthorized access but to protect from such challenges various methods and techniques are available which makes the smart grid network secure.

References

1. Ferrag MA, Maglaras LA, Janicke H et al (2018) A systematic review of data protection and privacy preservation schemes for smart grid communications. Sustain Cities Soc 38:806–835
2. Lopez C, Sargolzaei A, Santana H et al (2015) Smart grid cyber security: an overview of threats and countermeasures. J Energy Power Eng 9:632–647
3. Procopiou A, Komninos N (2015) Current and future threats framework in smart grid domain. In: 2015 IEEE international conference on cyber technology in automation, control, and intelligent systems (CYBER), pp 1852–1857
4. Lima C (2011) An architecture for the smart grid. In: Proceedings of IEEE P2030 smart grid communication architecture SGI ETSi workshop, pp 1–27
5. Kammerstetter M (2014) Architecture-driven SMART GRID security management. Presented at ACM workshop on information hiding and multimedia security
6. Gungor V, Sahin D, Kocak T, Ergut S, Buccella C, Cecati C, Hancke G (2013) A survey on smart grid potential applications and communication requirements. IEEE Trans Industr Inform 9(I):28–42
7. Huang A, Crow M, Heydt G, Zheng J, Dale S (2011) The future renewable electric energy delivery and management (FREEDM) systems: the energy internet. Proc IEEE 1:133–148
8. Sandberg H, Teixeira A, Johansson KH (2010) On security indices for state estimators in power networks. In: Preprints of the First Workshop on Secure Control Systems, CPSWEEK 2010
9. Dan G, Sandberg H (October 2010) Stealth attacks and protection schemes for state estimators in power systems. In: Proceedings of 1st IEEE SmartGridComm 2010, Gaithersburg, MD, pp 214–219
10. Overman TM, Sackman RW (October 2010) High assurance smart grid: smart grid control systems communications architecture. In: Proceedings of 1st IEEE SmartGridComm 2010, Gaithersburg, MD, pp 19–24

Intelligent Intrusion Detection Algorithm Based on Multi-Attack for Edge-Assisted Internet of Things

S. Shitharth, Gouse Baig Mohammed, Jayaraj Ramasamy, and R. Srivel

1 Introduction

The Internet of Things (IoT) connects multiple electronic devices with limited resources and enables a wide variety of uses. Users of the Internet of Things will frequently require access to cloud-based computing and storage services. In contrast, sending information from the Internet of Things to the cloud places additional strain on the network and causes delays in transmission. The impact of moving massive amounts of data to the cloud has been somewhat muted, however, due to the rise of edge computing as an efficient approach. According to Cisco, there will be 12.3 billion mobile connected devices in use by 2022, which is more than double the estimated global population of 8 billion in that year [1]. Industrial internet of things (IIoT) has played a significant role in several fields, including smart cities and industrial automation, due to the proliferation of IoT connections and

S. Shitharth (✉)
Department of Computer Science and Engineering, Kebri Dehar University, Kebri Dehar, Ethiopia
e-mail: shitharths@kdu.edu.et

G. B. Mohammed
Department of Computer Science and Engineering, Vardhaman College of Engineering, Hyderabad, India
e-mail: gousebaig@vardhaman.org

J. Ramasamy
Department of Information Technology, Faculty of Engineering & Technology, Botho University, Gaborone, Botswana
e-mail: jayaraj.ramasamy@bothouniversity.ac.bw

R. Srivel
Department of Computer Science Engineering, Adhiparasakthi Engineering College, Melmaruvathur, Tamil Nadu, India

applications. When it comes to realizing intelligent industrial operation, IIoT can link the devices used in industry to the Internet and rely on M2M communications [2]. By gathering data on a single server, however, a Traditional centralized cloud computing (CCC) center may make its distant services available to users. This method incurs substantial data transmission costs and falls short of the mark for certain low latency service and computation needs. In order to fulfil these needs, mobile edge computing (MEC) places a cloud computing platform at the edge of the radio access network (RAN), which significantly lessens the demands placed on the cloud data center's bandwidth and processing power [3, 4]. When it comes to IIoT applications and services, such as smart cities and autonomous vehicles, MEC is a crucial enabler. The computation, connectivity, and caching capabilities of edge servers make them ideal for use in IIoT applications, where they may be used to reduce latency and boost network performance.

Despite the many benefits of edge computing, studies on its security are still in their early stages. Intrusion detection, defensive strategies, and so on are only a few examples of the numerous places where advancement is needed [5, 6]. If the invasion isn't halted in time, it will cost businesses a fortune. The development of an intelligent intrusion prevention model in an edge environment is crucial for the protection of IIoT devices and data. An intrusion detection system (IDS) is crucial to any robust security strategy [7].

Intrusion detection looks to be a powerful active defence mechanism with a range of detection approaches [8]. Network-based intrusion detection and host-based intrusion detection are two types of intrusion detection [9] that may be distinguished by the data source they depend on. A network-based intrusion detection system is one of the approaches for seeing network threats a little bit sooner [10]. Modern network data, on the other hand, has not only more complicated but also more multidimensional properties. Traditional machine learning methods must manually retrieve a substantial number of features if working with high-dimensional data characteristics. The technique is hard and difficult, and the computation requested is vast, making it impossible to satisfy the precision and criteria in real time of IDS [11].

IDS relies heavily on its detecting methods. Intrusions may be detected reliably and promptly using a good intrusion detection algorithm. Examples of intrusion detection strategies are found in statistical detection and data mining [12]. Edge servers have a considerable challenge when trying to upload network data inside the region of monitoring in order to examine any anomalies that may have occurred. Because of this, researchers have attempted to employ machine learning to spot cyber-attacks [13]. Machine learning intrusion detection systems rely on an artificial intelligence system that can learn to distinguish between benign and malicious actions. The features of data have been the primary focus of several earlier studies on intrusion detection algorithms [14]. In contrast, due to single attack learning, they excel simply at determining whether or not there is any anomalous data and can identify such attacks with high accuracy at different times. When many types of attacks combine over time, they become indistinguishable to them. In addition, there are intrusion detection approaches that don't take into account the reality of limited resources in edge environments. As a result, edge-assisted IoT networks require study of a variety of low-overhead and high-efficiency learning algorithms for IDS. This research introduces a BP neural network module that establishes a robust feature space by dynamically learning the optimum features to protect against attackers. However, existing IDS cannot detect assaults that employ a variety of tactics. Therefore, RBF neural networks are proposed for analysis of attack fractions in each assault. As a result of our research and testing, we show that IDS is capable of identifying and preventing several types of cyberattacks.

The rest of the paper is structured as follows. Edge-assisted IDS are discussed in Sect. 2. Section 3 provides a comprehensive breakdown of IDS. The evaluation of the frameworks and the outcomes are then detailed in Sect. 4. The final observations are provided in Sect. 5. Finally, Sect. 6 concludes the research work with feature directions.

2 Related Work

Intrusion detection is an effective active defence strategy, and there are several intrusion detection techniques [15]. Network-based and host-based intrusion detection techniques are the two most common [16]. Using this strategy, a detection model is trained and tested using a pre-collected dataset annotated with normal and malicious behavior. Machine learning algorithms like the Bayesian model [17], support vector machine, and genetic algorithm have all seen extensive application in anomaly-based intrusion detection [18], in large part due to the impressive accuracy with which they execute in classification tasks. On the other hand, modern network data exhibit greater, more intricate, and multidimensional characteristics. Traditional machine learning approaches require a significant number of characteristics to be extracted manually when dealing with high-dimensional data. Complexity and computational size prevent it from satisfying intrusion detection's need for precision and timeliness [19].

The research on IDS for IIoT is extensive. In [20], the authors presented a decentralized intrusion detection method for IoT Network. Data mining is used to develop an ad hoc intrusion detection solution for the IIoT environment in [21], which focuses on detecting routing assaults.

Since the complexity and frequency of modern attacks are only expected to grow, it's important that we have access to cutting-edge tools for autonomous intrusion detection, such as machine learning. Some learning approaches may be used to increase the detection accuracy and speed of a NIDS that is powered by machine learning, allowing it to obtain the complicated aspects of the attack behavior. The Internet of Things (IoT) and machine learning are the subject of a few published works [22]. The nonparametric Bayesian Approach to network intrusion detection was first proposed in [23]. In [24], the authors proposed an intrusion detection system for computer networks that makes use of convolutional neural networks. The authors of [25] offer a deep belief network for intrusion detection. (DBN). In [26], the authors proposed a game-changing random Neural Network for detecting IoT network intrusions (RNN). The integration of RNNs with GAs improves the speed and precision of real-time decision making. To improve the speed and accuracy of intrusion detection, numerous reinforcement and multi-task learning techniques are provided in [27].

The scenarios presented in these publications, however, do not center on IDS in contexts with constrained hardware and software at the edge of the network. In contrast to IDSs in environments with no limits on processing power, those located at the network's periphery will encounter significant difficulties in protecting IoT systems. That is why research on lightweight and high-efficiency IDS is important in MEC. Additionally, an intrusion detection algorithm for single assaults can be used to uncover certain types of attacks. Few studies address the issue of attack prevention in the face of several simultaneous attacks on the edge of a network, or multi-attack scenario. The use of ANNs [28], SVM [29], k-nearest neighbour [30], random forest [31], deep learning techniques [32], Bayesian approaches, and decision trees are all examples of such methods. This has led to improved robustness in detection method performances.

To identify cyber threats, many organizations utilize artificial neural networks (ANNs), a computational paradigm that attempts to simulate the way the human brain's neural systems function. For the purpose of identifying DDoS/DoS assaults, Hodo et al. [19] describe a multi-level perceptron, a supervised ANN. The trial findings show that it is 99.4% accurate and can identify a wide range of DDoS/DoS assaults. For intrusion detection [33], offer a recurrent neural network-based deep learning method (RNN-IDS). The experimental outcomes demonstrate the efficacy of RNN-IDS in developing a highly accurate classification model. Using a deep convolutional neural network (DCNN) [34], present an intrusion detection system (IDS). Anomaly detection systems in the actual world can benefit from the proposed DCNN-based IDS, as demonstrated by the experimental findings. Cyber intrusion detection using an evolving neural network (ENN) that blends ANN with an evolutionary algorithm is demonstrated experimentally in [35]. This is a method used in research where the hyperparameters are systematically tweaked until the best possible answer is obtained. In [36], Shenfield et al. offer an innovative method for identifying malicious cyber traffic with artificial neural networks well-suited for application in deep packet inspection-based IDS. The findings provided here demonstrate that our innovative classification strategy can detect shell code with a very low false positive rate. Using ANN [37], offer a DoS attack detection method for wired local area networks [38, 39]. In their training dataset, the suggested ANN classifier achieves an accuracy of 96%. The categorization accuracy of most methods is encouraging.

Like the human brain, ANN can learn from examples and approximate the mapping of arbitrary functions. For the purpose of classification, ANN often outperforms the more traditional statistical approach [40]. However, creating an ANN from scratch is no easy task. Success depends on optimizing a large number of design elements, such as the optimal number of hidden nodes, learning algorithm, learning rate, and initial weight value, to achieve goals that are in tension with one another, including accuracy and complexity. This is why, when developing ANN, multi-objective optimization (MOO) is preferred over the more traditional single-objective approach [41]. However, ANN isn't perfect; it sometimes converges slowly, gets stuck on local maxima, and has an unreliable network architecture [42]. Contrarily, the genetic algorithm (GA) is the most popular method for mining data and uncovering hidden insights because of its global search and rapid convergence capabilities [43]. Conversely, Pareto optimality is a popular strategy in MOO [44]. Instead of providing only one best option, it presents a set of viable alternatives from which to choose the most effective course of action (individual and collective solutions).

3 System Model

Feature selection, dataset learning, and intrusion detection are the three primary components of the proposed edge-assisted IDS, as illustrated in Fig. 1. Features should be selected and trained before deploying intrusion detection. Data-based intrusion detection and time-interval based intrusion detection are the two phases that make up the proposed edge-assisted IDS. The BP neural network module is used by data-based intrusion detection schemes to identify data that deviates from regular patterns. The RBF neural network is used by the time interval-based intrusion detection technique to identify the proportion of attacks within a certain time frame of data collection. In specifically, the process of feature selection is the primary driver of feature capture and feature extraction. Particularly well-known in the field of intrusion detection is the KDD'99 network assault dataset. Denial of Service (DoS), Probe, User to Root (U2R), and Remote to Local (R2L) are only four of the assaults included in the dataset's 41 total characteristics [31]. Here are the specifics of the four assaults that comprise this dataset:

In a denial-of-service (DoS) attack, the attacker assaults the system with requests over a short period of time. As a result, all of the available resources have been

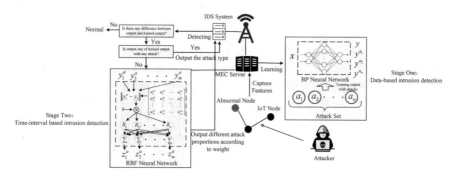

Fig. 1 Edge-assisted intrusion detection system

depleted, rendering it unable to carry out valid requests. Another type of Denial-of-Service attack is a distributed Denial of Service (DDoS) assault, in which the (1) attackers use many connections to force the service they are attacking to cease responding to users entirely or even crash. (2) By using a probe, an attacker can learn about a network's open ports, hosts, operating services, and more. (3) If an attacker has physical access to the system they are attacking, all they need to do to launch a User-to-Root (U2R) assault is exploit a vulnerability that allows them to gain access levels normally reserved for administrators. Buffer overflow attacks are a typical form of user-to-root (U2R) compromise, in which the attacker assumes the role of the system administrator in order to deal with malicious code. (4) Remote-to-Local (R2L) attacks occur when a hacker, who does not already have an account on the target system, pretends to be a known and trusted user on the local network. Combining a remote to local attack with a U2R assault is common. An instance of an R2L attack is a Secure Shell (SSH) brute-force assault. As a result, it is essential to make sure that the characteristics chosen are acceptable for the various kinds of attacks. The suggested intrusion detection method won't work if the characteristics used aren't appropriate for the assault. Data-driven intrusion detection learning seeks to identify information that deviates from regular patterns. As data is generated at regular intervals, time interval-based intrusion detection may learn to identify the relative frequency of each assault. During testing, a test instance might be labelled as typical or exceptional based on what was learned in the previous step. Finally, intrusion detection is used to determine if the behavior seen during testing is typical or indicative of an assault(s).

4 Proposed Intrusion Detection

A. Feature selection
The significance of feature selection [32] has been examined by a few publications utilizing a learning technique. Each feature's value is shuffled about randomly in those learning methods. The datasets are trained using a machine learning model. Using the rise in classification error as a metric, we evaluate the relative significance of the features. Changed features are seen to be significant if they lead to a large rise in classification error, whereas those that lead to a little increase in error are deemed to be unimportant. According to [33], Random Forest (RF) may be used to determine the significance of features by training on vulnerable KDD'99 datasets. For a lighter load on the computer. Figure 2 displays the top five characteristics. The greater the difference between normal and attack production during training, the more vital this quality becomes. The percentage of packets that had to be resent or dropped is represented by the total percent loss (Ploss), the percentage of packets that originated from the source port is represented by the source loss (Srcloss) metric, and the number of the port to which the packets were delivered is represented by the destination port (Dport) metric.

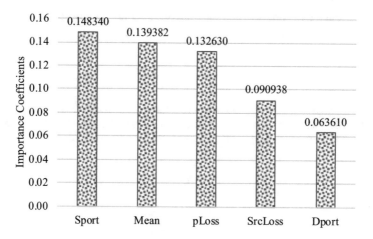

Fig. 2 The top five most important features

Fig. 3 Visualization of the inner workings of a neuron in a neural network

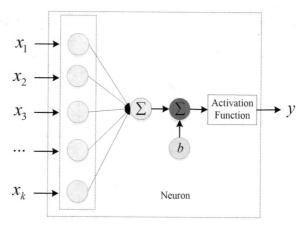

B. Dataset learning using BP and RBF neural networks

In view of the limited resources of edge servers, we provide lightweight BP neural networks to mimic input x and output y during normal and attack scenarios. A neuron is a computational unit defined by its inputs (x_1, x_2, ... x_k), their associated weights (w), and the threshold bias (b) (see Fig. 3). Finding the optimal values for the network's parameters is the aim of BP neural networks (typically its weights w and biases b). The outcomes might be written as.

$$y = f\left(\sum_i w_i x_i + b_i\right) \tag{1}$$

where y_i is the result variable. The cost function is defined as the difference between the dth estimated output y_d and the dth observed output y_d':

$$Z_d = \frac{1}{2} \| y_d - y_d' \|^2 \tag{2}$$

In order to reduce the loss function, we employ a gradient descent approach. To minimize the loss, the network parameters are iterated upon until convergence is achieved. The following adjustments will be made to the weights and biases:

$$w^* = w - \eta \frac{\partial Z}{\partial w} \tag{3}$$

$$b^* = b - \eta \frac{\partial Z}{\partial b} \tag{4}$$

where η is the rate of learning. The optimal values of weights w and biases b can be determined by a predetermined number of repetitions. Feature selection, meanwhile, is flexible and may be made based on how critical they are in stopping a given type of attack. However, it's possible that certain outcomes result from a combination of many forms of attacks targeting the edge network at the same time. In contrast to learning from a single assault, learning from many attacks is unable to identify mixed attacks. In order to amuse ourselves, we may define an attack set $A = \{a_1, a_2, \ldots, a_n\}$, and the trained output of each assault is $y^a = \{ y^{a_1}, y^{a_2}, \ldots, y^{a_n} \}$. Various assaults might affect the output in a way that makes it deviate from the expected value Z_d.

$$Z_d^a = \sum_{p=1}^{n} w_d^{a_p} y_d^{a_p} \tag{5}$$

A high-performance RBF neural network is used to find the connection between each attack y_d^a and Z_d. RBF neural networks, like most others, have an input layer, a hidden layer, and an output layer. In Fig. 1, we see an example of a multi-input multi-output RBF neural network. Nodes in the hidden layer are often represented by a radial basis function, whereas those in the output layer are typically represented by a linear function. A hidden layer neuron's response is localized and signal-dependent because of how its activation function works. A Kernel Function with a rather central location is predicted in the input signal. To find the relationship between y_d^a and Z_d, we apply RBF neural networks to the estimated output of each attack y_d^a from BP neural networks, using the normal output from BP neural networks as the center. The output is defined as the negative of the standard deviation of the output, or Z_d. Typically, the Gaussian function is used as the RBF neural network's radial basis function, and this is because of the following:

$$R_d\left(y_d^{a_p}\right) = \exp\left[\frac{-\mid\mid y_d^a - y_d \mid\mid}{2\sigma_d^2}\right], \qquad d = 1, 2, \ldots, m \tag{6}$$

where R_d is the radial basis function, y_d is the Gaussian function's central value, and represents the function's length. Therefore, an RBF neural network's input and output are mapped to one other as:

$$Z = \sum_{d=1}^{m}\sum_{p=1}^{n} w_m^{a_p} R_d\left(y_d^{a_p}\right) \tag{7}$$

A cost function is defined in the same way as the BP. It is the difference between the dth estimated output Z_d and the dth observed output Z_d'.

$$C_d = \frac{1}{2}\left\| Z_d - Z_d' \right\|^2 \tag{8}$$

In order to reduce the loss function, we employ a gradient descent approach. To minimize the loss, the network parameters are iterated upon until convergence is achieved. Here are the revised parameters for the weights and biases:

$$w^* = w - \lambda\frac{\partial C}{\partial w} \tag{9}$$

$$b^* = b - \lambda\frac{\partial C}{\partial b} \tag{10}$$

where λ is the rate of learning. Appropriate weights $w_m^{a_p}$ and biases b can be determined by a fixed number of repetitions. Consequently, the attack percentage for assault a_1 may be expressed as where $w_m^{a_p}$ is the weighting function.

$$\delta^{a_1} = \frac{\sum_{d=1}^{m} w_d^{a_1}}{\sum_{d=1}^{m}\sum_{p=1}^{n} w_d^{a_p}} \tag{11}$$

C. Intrusion detection

The trained BP neural network and RBF neural network are then used to forecast a testing packet's categorization during the intrusion detection step. The outcome is easily estimated by utilising the feature set b from the test packet in conjunction with the weight parameters in w acquired during training. The ideal weight parameter set w is stored following training of a BP neural network and an RBF neural network. Once the network has received the input characteristic, it may assess whether or not the packet is normal based on the value of the output at the last node, which is either 0 or 1. Following is the procedure for using the system:

S. Shitharth et al.

1. First, we use the feature selection technique to get the essential starting feature data for our detection findings;
2. In order to create the groundwork for future improvements to the intrusion detection system, it is necessary to insert some sample feature data into the training library;
3. The training procedure for a BP neural network consists mostly of two steps: the first optimizes the connection weight value, while the second optimizes the learning rate, both of which are based on the data set selected from the training library;
4. The BP neural network's connection weight and learning rate may be refined using data from other sources, and the network can then be fine-tuned using the optimized weight and rate.

5 Results and Discussion

In this section, the KDD99 dataset is used to measure the performance. The KDD99 technique, created by MIT Lincoln Labs, is the industry standard for intrusion detection. The KDD99 dataset principally consists of 4 classes and 39 different varieties of attacks. The training set has 22 of these attack kinds, whereas the test set features 17 unknown attack types. The detection rate of an intrusion detector is typically used as a metric of its effectiveness. The following is a description of a few of the variables:

Ture Negative (TN): shows the number of normal packets that have been correctly split into normal.

Ture Positive (TP): shows the percentage of suspicious data packets that can be successfully classified as attacks.

False Positive (FP): means the total number of benign packets that were mistakenly classified as malicious.

False Negative (FN): refers to the amount of misclassified aberrant packets.

The following is an evaluation of performance based on the aforementioned criteria:

The accuracy of a forecast is measured as the proportion of instances where the anticipated value is true.

$$Accuracy = \frac{TP + TN}{TP + TN + FP + FN} \times 100 \qquad (12)$$

The Undetected Rate (UR) is the percentage of suspicious data that passes undetected.

$$UR = \frac{FN}{FN + TP} \times 100 \qquad (13)$$

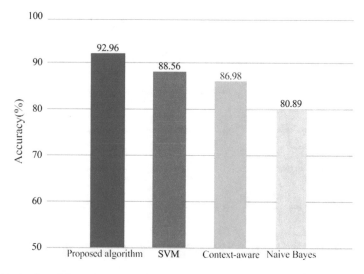

Fig. 4 Evaluation of the performance of several intrusion-detection algorithms

The fraction of benign packets incorrectly identified as malicious is known as the False Alarm Rate (FAR).

$$FAR = \frac{FP}{FP + TN} \times 100 \tag{14}$$

We test how well our system does at detecting intrusions based on data. To demonstrate the effectiveness of our proposed model, we evaluate it against three other methods: a support vector machine (SVM)-based method [34], a context-aware method [35], and a naïve bayes method [36].

Accuracy for a number of IDS techniques is displayed in Fig. 4. The suggested IDS has the maximum accuracy, whereas the naïve bayes detection method has the lowest accuracy. This is because the suggested BP neural network-based intrusion detection algorithms may dynamically learn the most effective characteristics to defend against intruders. Furthermore, the suggested IDS is able to identify mixed assaults that result from many vectors of attack. The suggested IDS outperforms the SVM based detection strategy, the context-aware detection approach, and the naïve bayes detection approach in simulations.

The false alarm rate (FAR), shown in Fig. 5, is the proportion of benign data packets incorrectly labelled as malicious ones. All detection strategies, with the exception of the naïve bayes detection strategy, exhibit promising results in Fig. 5. This is due to the inability of naïve bayes intrusion detection algorithms to respond appropriately to changing conditions. Moreover, naive bayes fails to detect attacks that combine techniques. Compared to SVM-based detection, context-aware

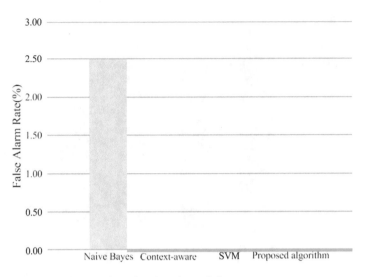

Fig. 5 FAR analysis of various intrusion-detection techniques

detection, and naive bayes detection, simulation results show the proposed IDS can achieve lower FAR. As can be seen in Fig. 6, the UR represents the percentage of anomaly packets that were incorrectly labelled as normal. Comparisons to different detection methods show that the proposed IDS has the lowest UR. This is so because the suggested intrusion detection algorithms can tell the difference between a single attack and a mixed assault that combines several techniques. The proposed IDS has been shown to outperform SVM-based detection, context-aware detection, and naive bayes detection in simulations. Under the total amount of concealed nodes, the proportion of the total mistake is shown in Fig. 7. Neural networks can only function optimally if they have the right kind of hidden nodes. Remember that false FAR is the proportion of benign packets wrongly identified as malicious ones. Although the FAR drops as the number of concealed nodes grows, it stabilizes at a certain point. It shows that increasing the number of concealed nodes does not improve accuracy. Depending on the context, a different number of hidden nodes may be required of neural networks deployed in the IoT.

Figure 7 shows how the proposed detection algorithm and the comparison method use CNN to provide four distinct URs for the same set of assaults. Using the two learning-based detection algorithms, we see decreased URs for DoS, probe, and higher URs for R2L and U2R. The suggested detection technique also has lower URs than the competing approach, which makes use of a convolutional neural network (CNN). The reason behind this is that when comparing algorithms that use CNN, only the case of a single assault to learn from is taken into account. Algorithm comparisons based on CNN only detect a single assault when an IoT network is subjected to a mixed attack. As a result, several opportunities to strike

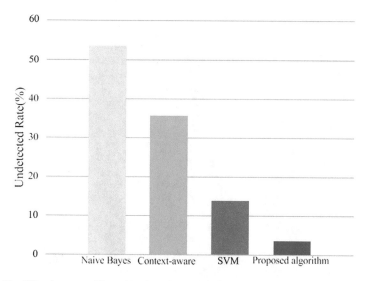

Fig. 6 The UR using several intrusion detection techniques

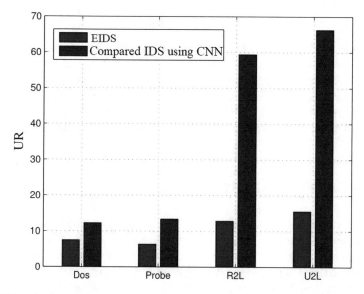

Fig. 7 UR under four attacks

were lost. The experimental outcomes verify the efficacy of the suggested technique in a mixed assault scenario.

In Fig. 8, we can see how the proposed detection algorithm and the comparative approach fared in detecting four distinct assaults using CNN. As can be shown, the FARs of DoS, probe is smaller when employing the two learning-based detection method, however the FARs of R2L and U2R are larger. In addition, the FARs of the

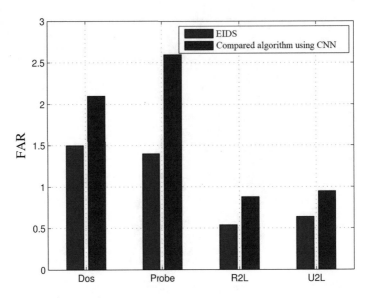

Fig. 8 FAR under four attacks

suggested detection method are smaller than those of the comparative approach that makes use of CNN. The reason for this is because while the comparable method utilizing CNN may have high performance in FARs under singer assault, it cannot match the detecting algorithms when they are in a mixed attack. Consequently, the suggested detection method outperforms the comparable approach when measured in FARs. The comparative technique makes use of a convolutional neural network.

6   Conclusion and Future Work

In this research, we show a multi-attack IDS that is supported by the mobile edge to deal with the problems of finding anomalies in the IoT. It uses BP neural network-based features to improve the accuracy with which unusual patterns can be found. To get beyond MEC's processing capability limitation and improve the training data trail, the recommended RBP neural network is used in combination with the mobile edge. In addition, the suggested technique may be applied to other forms of attack detection. According to the findings of the evaluation, the suggested technique is accurate to within a percent. Through the use of mobile edge. We plan to investigate further vulnerabilities in MEC servers as part of our future work.

References

1. Mohammad GB, Shitharth S, Syed SA, Dugyala R et al (2022) Mechanism of Internet of Things (IoT) integrated with radio frequency identification (RFID) technology for healthcare system. Math Prob Eng, Hindawi. https://doi.org/10.1155/2022/4167700
2. Wu D, Yan J, Wang H, Wang R (2019) Multiattack intrusion detection algorithm for edge-assisted internet of things. In: 2019 IEEE international conference on industrial Internet (ICII), pp 210–218
3. Shitharth S, Kshirsagar PR, Balachandran PK, Alyoubi KH, Khadidos AO (2022) An innovative perceptual pigeon galvanized optimization (PPGO) based likelihood Naïve Bayes (LNB) classification approach for network intrusion detection system. IEEE Access 10:46424–46441. https://doi.org/10.1109/ACCESS.2022.3171660
4. Siriwardhana Y, Porambage P, Liyanage M, Ylianttila M (2021) A survey on mobile augmented reality with 5G mobile edge computing: architectures, applications, and technical aspects. IEEE Commun Surv Tutor 23(2):1160–1192
5. Khan WZ, Ahmed E, Hakak S, Yaqoob I, Ahmed A (2019) Edge computing: a survey. Future Gener Comput Syst 97:219–235
6. Abbas N, Zhang Y, Taherkordi A, Skeie T (2017) Mobile edge computing: a survey. IEEE Internet Things J 5(1):450–465
7. Yoosuf MS, Muralidharan C, Shitharth S, Alghamdi M, Maray M, Rabie OB (2022) FogDedupe: a fog-centric deduplication approach using multikey homomorphic encryption technique. J Sens, Hindawi. https://doi.org/10.1155/2022/6759875
8. Sabella D, Vaillant A, Kuure P, Rauschenbach U, Giust F (2016) Mobile-edge computing architecture: the role of MEC in the Internet of Things. IEEE Consum Electron Mag 5(4):84–91
9. Mohammad GB, Shitharth S, Kumar PR (2021) Integrated machine learning model for an URL phishing detection. Int J Grid. Distrib Comput 14(1):513–529
10. Furnell S (2004) Enemies within: the problem of insider attacks. Comput Fraud Secur 2004(7):6–11
11. Roman R, Lopez J, Mambo M (2018) Mobile edge computing, fog et al.: a survey and analysis of security threats and challenges. Future Gener Comput Syst 78:680–698
12. Almogren AS (2020) Intrusion detection in edge-of-things computing. J Parallel Distrib Comput 137:259–265
13. Khadidos AO, Khadidos AO, Manoharan H, Alyoubi KH, Alshareef AM, Shitharth S (2022) Integrating industrial appliances for security enhancement in data point using SCADA networks with learning algorithm. Int Trans Electr Energy Syst 2022:8457116, 17 pages. https://doi.org/10.1155/2022/8685235
14. Prasanth SK, Shitharth S, Praveen Kumar B, Subedha V, Sangeetha K (2022) Optimal feature selection based on evolutionary algorithm for intrusion detection. SN Comput Sci. https://doi.org/10.1007/s42979-022-01325-4
15. Abid Salih A, Mohsin AA (2021) Evaluation of classification algorithms for intrusion detection system: a review. J Soft Comput Data Min 2(1):31–40
16. Singh BN, Khari M (2021) A survey on hybrid intrusion detection techniques. In: Research in intelligent and computing in engineering. Springer, Berlin, pp 815–825
17. Selvarajan S, Manoharan H, Hasanin T, Alsini R, Uddin M, Shorfuzzaman M, Alsufyani A (2022) Biomedical signals for healthcare using Hadoop infrastructure with artificial intelligence and fuzzy logic interpretation. Appl Sci, MDPI. https://doi.org/10.3390/app12105097
18. Roman R, Lopez J, Mambo M (2018) Mobile edge computing, Fog et al.: a survey and analysis of security threats and challenges. Futur Gener Comput Syst 78:680–698
19. Mohammad GB, Shitharth S (2021) Wireless sensor network and IoT based systems for healthcare application. Mater Today Proc, Elsevier. https://doi.org/10.1016/j.matpr.2020.11.801
20. Raponi S, Caprolu M, Pietro RD (2019) Intrusion detection at the network edge: solutions, limitations, and future directions. In: Zhang T, Wei J, Zhang LJ (eds) Edge computing – EDGE 2019, pp 59–75

21. Khadidos AO, Shitharth S, Khadidos AO, Sangeetha K, Alyoubi KH (2022) Healthcare data security using IoT sensors based on random hashing mechanism. J Sens 2022:8457116, 17 pages. https://doi.org/10.1155/2022/8457116

22. Sharma R, Chan CA, Leckie C (2020) Evaluation of centralised vs distributed collaborative intrusion detection systems in multi-access edge computing. In: IFIP networking 2020

23. Divakaran JS, Prashanth SK, Mohammad GB, Shitharth D, Mohanty SN, Arvind C, Srihari K, Abdullah RY, Sundramurthy VP (2022) Improved handover authentication in fifth-generation communication networks using fuzzy evolutionary optimisation with nano core elements in mobile healthcare applications. J Healthcare Eng, Hindawi. https://doi.org/10.1155/2022/2500377

24. Liu L, De Vel O, Han Q-L, Zhang J, Xiang Y (2018) Detecting and preventing cyber insider threats: a survey. IEEE Commun Surv Tutorials 20(2):1397–1417

25. Panda SK, Mohammad GB, Mohanty SN, Sahoo S (2021) Smart contract-based land registry system to reduce frauds and time delay. Secur Priv 4(5):e172

26. Mohammad GB, Shitharth S (2021) Wireless sensor network and IoT based systems for healthcare application. Mater Today Proc:1–8

27. Dugyala R, Reddy NH, Maheswari VU, Mohammad GB, Alenezi F, Polat K (2022) Analysis of malware detection and signature generation using a novel hybrid approach. Math Probl Eng 22(1):1–13

28. Sudqi Khater B, Abdul Wahab AWB, Idris MYIB, Abdulla Hussain M, Ahmed Ibrahim A (2019) A lightweight perceptron-based intrusion detection system for fog computing. Appl Sci 9(1):178

29. Tian Z, Luo C, Qiu J, Xiaojiang D, Guizani M (2019) A distributed deep learning system for web attack detection on edge devices. IEEE Trans Industr Inform 16(3):1963–1971

30. Kohli V, Chougule A, Chamola V, Yu FR (2022) MbRE IDS: an AI and edge computing empowered framework for securing intelligent transportation systems. In: IEEE INFOCOM 2022 – IEEE conference on computer communications workshops (INFOCOM WKSHPS), pp 1–6

31. Cheng Y, Lu J, Niyato D, Lyu B, Kang J, Zhu S (2022) Federated transfer learning with client selection for intrusion detection in mobile edge computing. IEEE Commun Lett 26(3):552–556

32. Li B, Wu Y, Song J, Lu R, Li T, Zhao L (2021) DeepFed: federated deep learning for intrusion detection in industrial cyber–physical systems. IEEE Trans Industr Inform 17(8):5615–5624

33. Geiping J, Bauermeister H, Dröge H, Moeller M (2019) Inverting gradients–how easy is it to break privacy in federated learning? Proc Neural Inf Process Syst (NeurIPS):1–11

34. Mugunthan V, Polychroniadou A, Byrd D, Balch TH (2019) SMPAI: Secure multi-party computation for federated learning. Proc Neural Inf Process Syst (NeurIPS):1–9

35. Chen Y, Qin X, Wang J, Yu C, Gao W (2020) Fedhealth: a federated transfer learning framework for wearable healthcare. IEEE Intell Syst 35(4):83–93

36. Yang H, He H, Zhang W, Cao X (Apr. 2021) FedSteg: a federated transfer learning framework for secure image steganalysis. IEEE Trans Netw Sci Eng 8(2):1084–1094

37. Xiao L, Lu X, Xu T, Wan X, Ji W, Zhang Y (2020) Reinforcement learning-based mobile offloading for edge computing against jamming and interference. IEEE Trans Commun 68(10):6114–6126

38. Chen M et al (2021) Distributed learning in wireless networks: recent progress and future challenges. IEEE J Sel Areas Commun 39(12):3579–3605

39. Mohammed I et al (2021) Budgeted online selection of candidate IoT clients to participate in federated learning. IEEE Internet Things J 8(7):5938–5952

40. Wang H, Kaplan Z, Niu D, Li B (2020) Optimizing federated learning on non-IID data with reinforcement learning. Proc IEEE Conf Comput Commun (INFOCOM):1698–1707

41. Chakraborty A, Misra S, Mondal A, Obaidat MS (2020) Sensorch: Qos-aware resource orchestration for provisioning sensors-as-a-service. In: ICC 2020 – 2020 IEEE international conference on communications (ICC), pp 1–6

42. Loukas G, Vuong T, Heartfield R, Sakellari G, Yoon Y, Gan D (2017) Cloud-based cyber-physical intrusion detection for vehicles using deep learning. IEEE Access 6:3491–3508
43. Tesei A, Luise M, Pagano P, Ferreira J (2021) Secure multi-access edge computing assisted maneuver control for autonomous vehicles. In: 2021 IEEE 93rd vehicular technology conference (VTC2021-Spring), pp 1–6
44. Alladi T, Kohli V, Chamola V, Yu FR (2021) Securing the internet of vehicles: a deep learning-based classification framework. IEEE Netw Lett 3(2):94–97

Secure Data Analysis and Data Privacy

Okash Abdi, Gautam Srivastava, and Jerry Chun-Wei Lin

1 Introduction

Traditionally, organizations have considered data a productive asset in their daily operations. Enterprises acquire more and more data whenever the chances arise. Consequently, they end up with too much data, which is challenging to manage and protect. Thus, organizations might have sensitive information ending up in malicious hands. Technological advancements have developed more uncomplicated hacking techniques requiring organizations to be more watchful over their data than in past times. Traditional data protection methods are gradually becoming ineffective in current computing environments. Organizations have adopted advanced data security methods, tools, and strategies to keep off invaders and prevent data loss. Data security analysts must continuously monitor data trends and activities in their organizations. The analysts should emphasize each process involved in data management: data collection, preparation, analysis, storage, and usage [11].

Additionally, analysts must conduct regular security audits on data and systems. One current system, the Internet of Things (IoT), allows easy access to global databases through networks of heterogeneous devices, sensors, and actuators. Enterprises must thus invest in securing all sensitive data connected through IoT. The process is costly and might interrupt the regular flow of other business activities. Thus, many organizations have moved their data to the cloud. Currently, the

O. Abdi · G. Srivastava (✉)
Department of Mathematics and Computer Science, Brandon University, Brandon, MB, Canada
e-mail: ABDIOM50@brandonu.ca; srivastavag@brandonu.ca

J. C.-W. Lin
Department of Computer Science, Electrical Engineering and Mathematical Sciences, Western Norway University of Applied Sciences, Bergen, Norway
e-mail: jerrylin@ieee.org

© The Author(s), under exclusive license to Springer Nature Switzerland AG 2023   137
G. Srivastava et al. (eds.), *Security and Risk Analysis for Intelligent Edge Computing*, Advances in Information Security 103,
https://doi.org/10.1007/978-3-031-28150-1_7

primary issue is determining which cloud provider offers the most benefits without tampering with the data [19]. The current trend in Artificial Intelligence (AI), deployment of cognitive AI in data security, has enabled securing data from a human perspective [15]. The development of quantum computing technology, in theory, promises a near-perfect environment, but it is still very much in the early stages of implementation.

2 Data Types

Data is made up of recording methods like numbers, characters, and images that can be analyzed and interpreted to make specific decisions. Data is divided into two types which are qualitative and quantitative. The two types have different strengths, and logic is often applied best to address different questions, situations, and/or purposes. Qualitative data refers to information represented in verbal or narrative format. Quantitative data refers to information expressed in large and small numerical terms corresponding to a specific category. Data undergoes a cycle involving analysis and processing to check the integrity and utilize the data. Data analysts take every step as significant as the others since they all contribute to the overall output. Any manipulation of the honesty of one process contaminates the whole cycle. It is thus mandatory for all enterprises to invest equally in managing all processes. Figure 1 shows the eight steps of the data cycle.

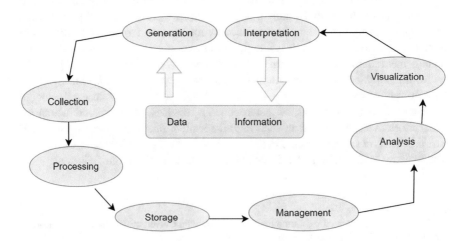

Fig. 1 Data cycle illustration

2.1 Data Analysis

The process of inspecting, cleansing, transforming, and modelling data to discover important information, and coming up with conclusions that support decision-making is known as data analysis. The process requires professionalism and keenness as it involves manipulating data without altering its meaning. Firms should assign trained personnel to the role of data analysts in their enterprises to avoid dependence on sheer luck. Enterprises employ different analysis strategies in data analysis. Visualizing involves creating a picture or graphic display of the data. Visualization is usually the starting point of analyzing data. The exploratory analysis includes creating a baseline for future analysis by looking at the data to identify or describe "happenings". The task is mainly done when there is a low level of knowledge about particular indicators. Trend Analysis entails going through data collected at different periods to detect change. Estimation works by using actual data values to predict future values over time. Estimation is used as a tool to plan for the future. Organizations or individuals interested in data analysis should invest in at least two strategies to complement each other. Many enterprises utilize all the named strategies and other peripherals to compete relevantly and constantly under different analytic and objective situations.

The type of analysis employed depends on the nature of the data and the objective questions aimed to be answered. The major types are descriptive analysis which is generally designed to answer the question "what happened", intending to summarize data in a meaningful manner without making predictions. Exploratory analysis, which goes further, skims data for available patterns and trends. There is also descriptive analysis and diagnostic analysis that involves using insights and information that are both descriptive and exploratory to investigate, find, and come up with related or actual causes. Predictive analysis uses data, statistics, and machine learning algorithms to predict future outcomes based on data. Prescriptive analysis utilizes all the other types of data to determine proper courses of action. Each type of analysis answers a specific objective question. Thus, data analysts end up utilizing all types at different stages. Effective data analysts should be well informed on all the types to protect their enterprise's data from interference. The flexibility to shift over different types determines the enterprise's ability to withstand different data attacks.

The types and strategies of data analysis require appropriate methods of implementation. Some of the major methods utilize Cluster Analysis involving organizing data in groups with similar characteristics. Regression Analysis involves sets of statistical processes that assist in examining the relationship between two or more variables. Factor Analysis compiles several variables into just a few, thus making analyzing data more manageable. Text Analysis involves extracting machine-readable information from unstructured texts such as documents and emails. Data Mining is used to trace trends, patterns, and correlations in large data sets for analysis. Analysts utilize several of these methods simultaneously or consecutively to acquire appropriate information (knowledge) from the data. The range of methods also increases the probability of sighting threats on time.

2.2 Methods of Analyzing Data and Data Analytic Tools

The tools used in data analytics are made up of various software applications commonly used in data analytics. The utilization of a tool depends on the nature of the data and the process of analysis. The major software active currently includes R-Programming which is a tool widely used for statistics and data modelling to easily manipulate data and, after that, present it in different ways. The tool is considered reliable in terms of the capacity of data, performance, and outcome. R-programming is well known for compiling and running on various Operating Systems, has many packages that can be browsed by categories, and provides tools to install all packages as required automatically. Python is a popular tool in computer programming. Guido Van Rossum developed the tool in the 1980s to support both functional and structural programming methods. The tool is based on an object scripting language that is easy to read, write and maintain data. Python is complemented by its learning libraries. The tool can be assembled on any platform like SQL Server JSON because of its ability to handle text data. Additionally, Excel is an analytical tool that is basic, widely used, and mostly considered essential in analyzing the client's internal data, breaking down the complex task with a preview of pivot tables, and later filtering it as per client requirements. Excel has advanced business analytics options that assist in modelling capabilities like automatic relationship detection, DAX measures, and time grouping.

Other software tools include Rapid Miner (RM), Tableau public, and SAS Enterprise Miner. Rapid miner is a powerful integrated data science platform that handles predictive analysis and advanced analytics like data mining, text analytics, machine learning, and visual analytics. Without any programming tool, RM links well with any data. The tool is unique in generating analytics based on natural–life data transformations. Industries and data analysts can thus control formats and data sets for predictive analysis. Tableau is software that connects to any data source and creates data visualizations, maps, and dashboards with constant updates through their web-based platform. Tablaeu allows access to download files in different formats and can hold large amount of data. It provides interfaces for analysis and visualization that are considered among the best. SAS Enterprise Miner is a tool that utilizes a programming platform and a language for data manipulation developed by the SAS institute. The tool is easily accessible and manageable, as well as can analyze data from any source. SAS is historically reckoned for introducing a large set of products for customer intelligence and numerous modules for web, social media, and marketing analytics in 2011.

Analysts prefer to use software they know well as they normally have systemic procedures. Having more than one software skill provides an edge over other data analysts. Data analysts are thus encouraged to expand their software skills scope

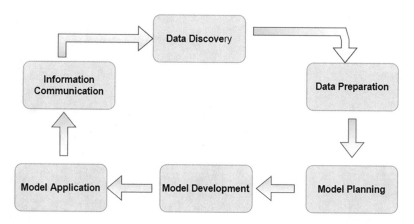

Fig. 2 Data analysis cycle

to be reliable at all times. Figure 2 illustrates the data analysis lifecycle involving preparing data for processing, developing an optimal analysis model, implementing the model, and communicating the results.

3 Data Security

Data security refers to the art of protecting data from intrusion, corruption, and loss. The process can be implemented through various strategies and methods as will be discussed in this section.

3.1 Types of Data Security

The primary data security tools utilized in current operations are encryption, data erasure, data masking, and data resiliency. Analysts should be conversant when to utilize different types of tools. Encryption refers to using algorithms to change standard text characters to unreadable format. The algorithms scramble the data so that attackers cannot deduce any information [10]. The algorithms protect data and develop security initiatives like authentication and integrity. Encryption is divided into several types, mainly asymmetric and symmetric encryption methods [25]. On the other hand, the asymmetric encryption method encrypts and decrypts data using two cryptographic asymmetric keys known as the public and private keys. RSA is an algorithm named after Ron Rivest, Adi Shamir, and Leonard Adleman that uses a public key to encrypt data and a private key to decrypt it. Public critical infrastructure is a means of managing encryption keys by providing and controlling

Fig. 3 Data security lifecycle

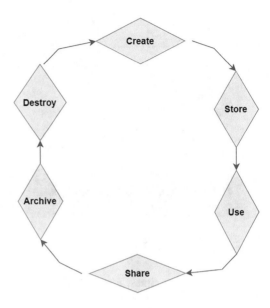

digital certificates. Symmetric encryption methods utilize only one secret symmetric key in encrypting the plaintext and decrypting the ciphertext. Twofish is an asymmetric encryption method that offers the quickest encryption algorithms, and it is freely available. Some of the other methods are Data Encryption Standard (DES), Triple DES (TDES), and the Advanced Encryption Standard (AES). DES is a low-level encryption algorithm that transforms plain texts in 64-bit blocks into ciphertext using 48-bit keys. TDES is achieved by running DES encryption three times successively. AES is a method used worldwide and is considered the US government standard.

3.2   Tasks in Securing Data

Tasks for securing data are the activities involved or associated with the data security process. Some activities act as prerequisites, while others develop as the process continues. The most efficient tasks are the ones whose vitality extends to the following security cycle. Figure 3 shows the primary tasks in a data security cycle.

There are data discovery and classification tools that automate identifying sensitive data, assessing it, and analyzing vulnerabilities. Not only is it critical to protect data, but also to correct and erase data upon demand. Data should be sorted appropriately to ease the processes of retrieval and usage. The technology used should be secure from maliciousness to maintain data integrity and efficiency for timely delivery.

Data and file activity monitoring which involves continuously monitoring activities to manage data to detect any unusual patterns that might imply a potential threat can be utilized. Instant alerting and blocking are utilized during unusual activity patterns to rectify and prevent threats altogether. The process allows for timely recognition of security threats and quick recovery.

There are tools such as vulnerability assessment and risk analysis which simplify identifying and assessing vulnerabilities such as weak passwords and out-of-date software as well as pointing out data sources with the most significant risk [14]. Sumo Logic, a machine data analytic service can be also adopted by organizations to supervise, assess, and rectify operational problems and security threats to their cloud services.

In Security Analytics, a combination of three open-source products are deployed, which include elastic search, Kibana, and log stash that are used by enterprises to identify and solve threats that are exposed to their systems. Automated compliance reporting can also provide a centralized repository where unit-wide data undergoes a typical audit trail. All data systems should be set to follow articulate measures. Central data repositories enhances system supervision and allow data analysts to spot anomalies easily. Analysts can implement changes or control over unit-wide systems from their central setting, thus saving time and cost.

4 Big Data Security Analytics

The responsibilities of IT security personnel are: (a) ensuring the company's systems are secure, (b) counterchecking cyber threat risks, and (c) adhering to data governance regulations [18]. Tasks tend to include spearheading the monitoring the process of analyzing large amounts of data from servers, network devices, and applications. Big Data Security Analytics is divided into two categories: Privileged Access Management (PAM) and Security Incident and Event Management (SIEM) [13].

PAM focuses on managing operational data [8] while SIEM systems bring together log data produced by monitored devices like computers to identify security-related events on individual machines. The tools emphasize log and event management, behavioural analysis, and finally, database and application monitoring.

The tasks, from knowledge discovery to big data analytics, must be effectively carried out for the output to be essential. Some enterprises may employ a single analyst for each process or a single analyst for all processes. The choice should be governed by workload. Every analyst should be able to comfortably handle their tasks efficiently if a firm is to benefit from their services.

4.1 Strategies for Data Security

Organizations implement diverse strategies in their data jurisdiction to keep it safe. The strategies should comprehensively cover the whole data system, including physical security. The most common strategies currently are physical security, data minimization, backup implementation, patch updates, and employee education in large enterprises.

Physical security is emphasized in all firms globally. Storage devices and facilities should be kept safe from the reach of intruders who might want to steal or destroy them. Access should be limited to relevant personnel only. Natural factors such as fire should be considered to provide optimal working environments and prevent data and material loss in the event of natural disasters. Firefighting equipment should be strategically positioned within the facilities.

The evolution of the Internet of Things (IoT) has led to more data-collecting means and invasion opportunities. Data minimization involves collecting just the relevant data necessary for a specific objective This trend leads to the minimization in the accumulation of junk data, which can complicate retrieving relevant data. Enterprises should thus implement data minimization policies to store essential data and delete the rest. Data minimization reduces the costs of data retrieval, storage as well as minimizes the chances for a significant leak of information, some of which might have been irrelevant to the organization [22].

To allow for easy supervision and tracking when security issues arise, access management and control strategies that involve limiting the network, and database access can be used. The authorized person(s) should be informed adequately on data security and potential threats before being cleared to access sensitive databases.

Once new patches are out, all software should be updated to the latest versions. Organizations should utilize the patches depending on their security vulnerability and the cyber threats they aim to resolve [4]. Effective patching should consider missing patches, faults during patching, change freeze periods, and feature patches in resource-constrained environments.

Maintaining backup copies of essential and reliable critical data is a must for any enterprise. All backups should be under the same physical and logical security controls. One of the best ways to safeguard data is Employee Education. Steps include training employees who interact with the data constantly, on the importance of data security, password privacy and to be familiar with policies and procedures in case a data breach occurs. Network and endpoint security monitoring and controls are essential in preventing external attacks The method involves using threat management, detection, and response tools to evaluate risks and prevent a breach on the cloud or on-premises environment.

Of the discussed strategies, none is of more significance than the other. The strategies should be utilized complementarily to block any loopholes in data systems. The fact that each strategy targets a particular objective in data security enhances the efficiency of utilizing several if not all the mentioned above. It is advisable to form a team to emphasize the implementation of the strategies.

4.2   Data Security Concepts

Here, we will discuss the current concepts associated with data security in business enterprises. The major concepts discussed are cloud storage, the use of personal computers, and the attainment of high-grade data. Data security and BYOD (bring your device) is a process that involves managing personal computers, mobile phones, and tablets used in enterprise computing environments. To improve security, the enterprise emphasizes installing security software to access shared networks, thus enabling centralized management over data access and movement.

Cloud-based infrastructure requires more complex models compared to traditional defences set at the network's perimeter [6]. Cloud monitoring involves exhaustive cloud data discovery and classification tools, continuous monitoring, and handling risks [2]. A good cloud provider excels in the following tasks:

- **Firm design and enterprise-grade performance**: the cloud should meet the client's requirements in performance, security, reliability, and sovereignty. The cloud should run on trusted servers and adequate infrastructure to isolate public internet and global data centers [7].
- **High availability and continuity**: a good cloud provider should have optimal uptime availability and offer resolutions in a failure. Site resiliency across several locations ensures the retention of data when the primary site is threatened. The cloud should utilize stretched clusters to prevent data loss during a complete failure [12].
- **Visibility, management, and consistency**: the cloud should allow clients to utilize the same tools and technologies as they would on their premises. The factor ensures maximum control and visibility [21]. Modernization capabilities: the cloud should allow for accessible developments and advancements, including building new applications and modernizing critical applications.
- **Security**: the cloud should have adequate security features to prevent malicious or accidental breaches. The process significantly requires the effective management of encryption keys. Role-based access control is essential in controlling and preventing configuration changes [16]. Smooth integration and automation: the cloud should allow the clients to leverage their existing tools, technologies, knowledge, and experience without sacrificing critical elements like security and resiliency during cloud migration [20].

4.3   Achieving Enterprise-Grade Data

The interpretation of enterprise-grade data is highly flexible depending on the field and context. The current trends have seen enterprises moving their operation applications to the cloud. The migration is a risky endeavour because a data breach or downtime experienced can lead to catastrophic losses. To commission the migration, IT teams should identify the cloud provider that will offer the highest levels of enterprise-grade performance.

Data analysts are currently keener on data security than in the past due to the rise in persistent threats. The rise in activities on this issue has led to data security associated with the cloud, BYOD or enterprise-grade data as the concepts have gained popularity coincidentally. The three concepts must thus be positively affected for the data security process to be deemed effective currently.

4.4 Methods of Secure Data Analytics

Data analysts utilize different methods to examine, prevent, and deal with data security issues. The methods are employed at various steps in the data security cycle singly or concurrently with each other. The most common methods are discussed extensively next:

Data mining, also known as Knowledge Discovery in Data (KDD), involves gathering data from internal databases and external sources for security data analysis. However, KDD generally means gathering data for research. The advancement of data warehousing has led to various data mining techniques. The techniques can describe the target data or involve mathematical and computational algorithms in recognizing various patterns in the extracted datasets. The process keeps threats away by monitoring potentially suspicious behaviour. Data cleaning assists in analyzing datasets and identifying the data relevant to security decisions. The process occurs in various steps, from collecting the data to visualizing it. The four main steps are (1) setting objectives, (2) data preparation, (3) utilizing data mining algorithms, and (4) evaluating the results [5]. Selecting the objectives is the hardest step though most organizations give this task little emphasis. Data scientists and stakeholders should be adequately informed of the context to set appropriate data questions and parameters of the process.

The relevant set of data is selected according to the objectives set. The data is then cleaned, and additional tools are used to narrow the features down to reduce bulk but still ensure optimal accuracy. Modelling and pattern mining require professional expertise to run properly. Data scientists investigate critical data relationships and highlight areas of potential fraud. The process utilizes frequency patterns and deviations. Deep learning algorithms can be utilized to cluster datasets. Supervised learning is used to group data when the input data is labelled, and unsupervised learning is used in unlabeled datasets.

On data aggregation, the results are evaluated and interpreted. The results should be simple, valid, and essential. Organizations utilize the knowledge acquired to develop strategies and objectives.

Data mining techniques are governed by the logical or statistical models employed. The major data mining techniques are:

- **Decision Tree**: utilizes regression methods to sort datasets or estimate potential outcomes. The technique employs a tree-like figure to illustrate the potential outcomes of various decisions.
- **Neural Networks**: imitates the human brain interconnection through nodal layers. The nodes consist of inputs and an output, and they pass data to adjacent layers in the network when the output exceeds a preset threshold.
- **Association Rules**: The technique utilizes rules to discover variable relationships in a dataset.
- **K-nearest Neighbour** (KNN): involves a non-parametric algorithm that sorts data sites according to their proximity and relationship with other data aiming to find the distance between data points on the assumption that similar data points are found nearby.

Data security is vital in analysis in ensuring honest outcomes. The process involves digging into the facts aiming to answer a single question. Analytical and logical reasoning is used to check data from security systems. Standard security analytics tools are used to deduce the data—the analysis of network traffic assists in noting trends that indicate a potential attack. Behavioural analytics is a popular concept in data analysis. The process involves checking trends of users and systems to recognize suspicious behaviour that could reflect a breach or potential security invasion. Real-time intrusion detection is mandatory in ensuring an efficient data security system. The process involves gathering data and conducting up-to-date analysis of security alerts from various sources, including devices and applications. Real-time intrusion detection detects malicious activities as they happen, thus assisting in a timely reaction to threats [23].

4.5   Security Audits

Security Audits involves conducting independent and orderly examinations of security systems to determine the validity of the existing information [9]. The audits assist in recognizing internal contamination of the data crucial in security decisions. When properly implemented, security audits recognize security gaps, improve security status, and confidence in the security controls, and enhanced technology and security performance [24]. For optimal efficiency, data forensic scientists utilize checklists in doing the examinations.

Forensic analysts focus on five major areas: (1) Data security, which reviews data protection methods like encryption, data accessibility, transmission, and storage security, (2) operational security, which examines regulations, policies, and procedures governing the data, (3) network security, which examines network controls, network security configurations, and monitoring capabilities, (4) physical security, that examines the physical devices like biometric disk servers and their functionality and finally (5) system security, which examines the effectiveness of various tasks in the system like patching processes and accounts management.

4.6 Forensics

Forensics involves preserving, recognizing, gathering, and documenting digital evidence from a compromised system. Forensic systems have various objectives, including assisting in retrieving and analyzing computers and related materials crucial for decision-making. Forensics is crucial to in identifying evidence quickly and correctly and estimate the effects of the malicious acts, while also aiding in the retrieval of deleted files and validating them, while also providing forensics reports showing the results of an investigation processes to assist in future remedies. Forensics includes various steps:

- **Identification**: involves recognizing and tracing the evidence present, its storage location, and its format.
- **Preservation**: involves obscuring people from utilizing the affected digital device to avoid the corruption of evidence. In this way, the isolated data is secured and preserved.
- **Analysis**: data is reconstructed, and conclusions are derived from the evidence. The process involves multiple examinations to support a specific theory.
- **Documentation**: involves recording all the visible data. The purpose is to recreate and review the attack process. The records utilize photographs, sketches, and security maps.
- **Presentation**: involves summarizing and explaining the conclusions of the investigation process in simple language. The summary will help develop strategies to prevent future threats and efficiently resolve them when they occur.

Digital forensics can be classified into three types: (1) disk forensics, (2) network forensics, and (3) database forensics. Disk forensics involves gathering data from active, modified, and deleted files from storage devices. Network forensics involves monitoring and analyzing network traffic to collect meaningful information and evidence. Wireless forensics is a sub-branch of network forensics that offers the tools to gather and process data from wireless network sources or traffic. Database forensics involves examining databases and their metadata. The three types can be classified into four sub-groups: (1) email forensics, (2) memory forensics, (3) malware forensics, and (4) mobile phone forensics.

The main challenges faced by the forensics process are the availability of practical hacking tools, the destruction of physical evidence, the need for large storage capacities, dynamic technological changes, and extensive use of internet access.

4.7 Visualization and Reporting

Visualization and reporting involves a data analyst presenting the results in an understandable form to stakeholders [1]. The techniques used are dependent on the type of data involved. Some common techniques include graphs, histograms, heatmap visualization, infographics, and fever charts. Graphs are excellent in illustrating the time-series relationship in a specific block of data. Histograms are graphical tools that arrange a set of data into a range of frequencies. Histograms arrange the information in a simple, understandable mode like a bar graph. Heatmap visualization is based on the human brain interpreting colours faster than numbers. The numerical data points are marked with different colours to show high or low-value points. Infographics: Infographics takes large amounts of data and arrange it for more straightforward interpretation. The tool is significantly effective when dealing with complex datasets. Fever charts display how data transforms over time.

Data analysis and visualization offer significant benefits in secure data analysis.

- **Improved decision-making**: when the right skills and software are used to analyze and illustrate data, enterprises can easily recognize the patterns and trends, thus making informed decisions on future endeavours.
- **Improved efficiency**: Proper data analytics and visualization show the enterprises the areas that require improvement and by what means to increase productivity.
- **Data relationships**: Analysis and visualization techniques assist in deducing the relationship between various data sets. When the enterprise notes the relationship between independent data sets, they develop strategies to maintain optimal performance. Improved insights: Visualization allows the division of data into smaller sets that assist in developing particular strategies to solve specific issues and understand the information effectively.

The methods are in fact steps in data analysis. The steps are implemented from a security perspective thus qualifying as methods of data security. It is the various segments in the primary methods named that secure the data. Security analysts are thus data analysts that are more concerned with the integrity of the data over preferable results. Being stages, the methods can be carried out by single personnel or a group of professionals depending on the workload.

4.8   Trends in Data Security

Trends in data security refers to the current advancements in the data security field involving current technologies utilized. Consequently, technologies are constantly developed to take on any arising security issues. AI and cloud technologies have taken bold steps towards a data secure environment while the fast-developing quantum computing aims at creating the perfect data security scenario. Artificial Intelligence amplifies the process of data security by providing the ability to handle large quantities. The capacity allows for efficient and timely decision-making when critically required. Cognitive AI, which reflects the human brain, is currently utilized to process data from a human perspective. The use of AI in Analysis enhances the validity of the process and ensures fast and timely resolutions [3].

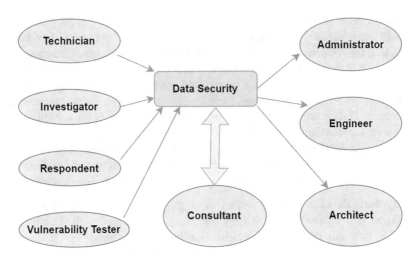

Fig. 4 Primary data security stakeholders

Multi-cloud security is a common trend globally. Currently, many enterprises have undergone cloud migration by moving their data and applications to the cloud. Cloud providers offer improved flexibility, easier scalability, faster development cycles, and improved customer experiences. A sound cloud must be secure, reliable, consistent, and high performing [2]. Similarly, Quantum computing is a groundbreaking evolution in computer science. Quantum algorithms allow for handling very complex data to find optimal outputs. Thus, quantum promises optimal security status and response.

Remarkably, security analysts are encouraged to be conversant with the current trends in data security or computing at large. The insight assists in identifying the benefits or shortcomings of the developing processors or systems. Thus, the three primary trends discussed can be beneficial or disastrous to a firm depending on the analysts' capabilities. Firms must invest in continuously informing their security analysts on upcoming issues either via offering the means or the training itself.

Figure 4 shows the primary data security stakeholders. The players on the left generate the data, the middle ones handle the process and those on the right implement results and solutions.

5   Data Privacy

Data privacy must be all enterprises' top priority. Sensitive data should be handled properly according to regulatory terms to preserve confidentiality. Since privacy is

Table 1 Three major methods of data protection

Traditional data protection	Data security	Data privacy
Archiving	Access control	Legislation
Physical infrastructure	Encryption	Global variations
Backup & recovery	Authentication	Policies
Data retention	Data loss prevention	Best practice
Replication	Threat monitoring	Data governance
Erasure coding	Breach access	3rd party contacts

mainly considered a legal concept, data privacy is controlled by legislation, policies, and global variations. Due to the current trends in cloud migration, many countries have developed legislation to prevent domestic data from leaving their countries. Thus, data privacy is mainly associated with geographical boundaries, while data protection defines technical security and data access. Data privacy is essential in managing the data asset in an organization and enhancing regulatory compliance. While data security protects data from internal and external threats, data privacy manages data collection, sharing, and usage modes [17]. Though different, data privacy should be taken as significantly as data security. Privacy improves people's confidence and trust in firms. Thus, whether governed by morality or legislation, enterprises must provide data privacy services to their dependents. Action, like data security, requires professionalism and objectivity to effectively occur. Table 1 compares the three major methods of data protection by outlining major activities in the methods.

6 Conclusion

Data is an essential asset for any enterprise. Traditionally, organizations favoured storing massive data amounts to have a vast data pool when called upon. As data size increases, their management and retrieval become more complicated. A lot of data is also tiresome to protect from being accessed by unauthorized individuals, theft, and corruption by internal and external sources. Security data analysts employ various tools, methods, and strategies to maintain data security. The methods are utilized according to the type of data involved and the specific objectives of processes. The methods can be used concurrently for an overall solution, or individually for more specific needs. Tools used should be updated to the current technological advancements and trends for optimal performance. Data privacy ensures that the legislative aspect of data protection is adhered to. The continuous evolution of AI, cloud and quantum computing will ensure better security and analytic tools.

References

1. Bao Y, Tang Z, Li H, Zhang Y (2018) Computer vision and deep learning-based data anomaly detection method for structural health monitoring. Struct Health Monit 18(2):401–421. https://doi.org/10.1177/1475921718757405
2. Behl A, Behl K (2012) An analysis of cloud computing security issues. In: 2012 World congress on information and communication technologies. https://doi.org/10.1109/wict.2012.6409059
3. Bertino E, Kantarcioglu M, Akcora CG, Samtani S, Mittal S, Gupta M (2021) AI for security and security for AI. In: Proceedings of the eleventh ACM conference on data and application security and privacy. https://doi.org/10.1145/3422337.3450357
4. Buck B, Hollingsworth JK (2000) An API for runtime code patching. Int J High Perform Comput Appl 14(4):317–329. https://doi.org/10.1177/109434200001400404
5. Chakarverti M, Sharma N, Divivedi RR (2019) Prediction analysis techniques of data mining: a review. SSRN Electron J. https://doi.org/10.2139/ssrn.3350303
6. Chen D, Zhao H (2012) Data security and privacy protection issues in cloud computing. In: 2012 International conference on computer science and electronics engineering. https://doi.org/10.1109/iccsee.2012.193
7. Chen Y (2020) IoT, cloud, big data and AI in interdisciplinary domains. Simul Model Pract Theory 102:102070. https://doi.org/10.1016/j.simpat.2020.102070
8. Dilmaghani S, Brust MR, Danoy G, Cassagnes N, Pecero J, Bouvry P (2019) Privacy and security of big data in AI systems: a research and standards perspective. In: 2019 IEEE international conference on big data (big data). https://doi.org/10.1109/bigdata47090.2019.9006283
9. Galdon Clavell G, Martín Zamorano M, Castillo C, Smith O, Matic A (2020) Auditing algorithms. In: Proceedings of the AAAI/ACM conference on AI, ethics, and society. https://doi.org/10.1145/3375627.3375852
10. Guo L, Xie H, Li Y (2020) Data encryption based blockchain and privacy-preserving mechanisms towards big data. J Vis Commun Image Represent 70:102741. https://doi.org/10.1016/j.jvcir.2019.102741
11. Gupta B, Agrawal D, Yamaguchi S (2016) Handbook of research on modern cryptographic solutions for computer and cyber security. IGI Global
12. Kandukuri BR, Paturi VR, Rakshit A (2009) Cloud security issues. In: 2009 IEEE international conference on services computing. https://doi.org/10.1109/scc.2009.84
13. Kantarcioglu M, Shaon F (2019) Securing big data in the age of AI. In: 2019 First IEEE international conference on trust, privacy and security in intelligent systems and applications (TPS-ISA). https://doi.org/10.1109/tps-isa48467.2019.00035
14. Kim D (2014) Vulnerability analysis for industrial control system cyber security. J Korea Inst Electron Commun Sci 9(1):137–142. https://doi.org/10.13067/jkiecs.2014.9.1.137
15. Klašnja-Milićević A, Ivanović M, Budimac Z (2017) Data science in education: big data and learning analytics. Comput Appl Eng Educ 25(6):1066–1078. https://doi.org/10.1002/cae.21844
16. Kumarage H, Khalil I, Alabdulatif A, Tari Z, Yi X (2016) Secure data analytics for cloud-integrated Internet of things applications. IEEE Cloud Comput 3(2):46–56. https://doi.org/10.1109/mcc.2016.30
17. Marin G (2005) Network security basics. IEEE Secur Priv Mag 3(6):68–72. https://doi.org/10.1109/msp.2005.153
18. Martinez LS (2017) Data science. In: Encyclopedia of big data, pp 1–4. https://doi.org/10.1007/978-3-319-32001-4_60-1
19. Mathisen E (2011) Security challenges and solutions in cloud computing. In: 5th IEEE international conference on digital ecosystems and technologies (IEEE DEST 2011). https://doi.org/10.1109/dest.2011.5936627
20. Pal AS (2013) Classification of virtualization environment for cloud computing. Indian J Sci Technol 6(1):1–7. https://doi.org/10.17485/ijst/2013/v6i1.21

21. Russell G, Macfarlane R (2012) Security issues of a publicly accessible cloud computing infrastructure. In: 2012 IEEE 11th international conference on trust, security and privacy in computing and communications. https://doi.org/10.1109/trustcom.2012.259
22. Senarath A, Arachchilage NA (2019) A data minimization model for embedding privacy into software systems. Comput Secur 87:101605. https://doi.org/10.1016/j.cose.2019.101605
23. Tan Z, Nagar UT, He X, Nanda P, Liu RP, Wang S, Hu J (2014) Enhancing big data security with collaborative intrusion detection. IEEE Cloud Comput 1(3):27–33. https://doi.org/10.1109/mcc.2014.53
24. Wicaksono A, Laurens S, Novianti E (2018) Impact analysis of computer assisted audit techniques utilization on internal auditor performance. In: 2018 International conference on information management and technology (ICIMTech). https://doi.org/10.1109/icimtech.2018.8528198
25. Xiong J, Chen L, Bhuiyan MZ, Cao C, Wang M, Luo E, Liu X (2020) A secure data deletion scheme for IoT devices through key derivation encryption and data analysis. Futur Gener Comput Syst 111:741–753. https://doi.org/10.1016/j.future.2019.10.017

A Novel Trust Evaluation and Reputation Data Management Based Security System Model for Mobile Edge Computing Network

Gouse Baig Mohammed, S. Shitharth, and G. Sucharitha

1 Introduction

The cloud computing platform gives users access to a broad variety of services and a nearly endless supply of resources, which enables them to take advantage of a number of different opportunities [1, 2]. The availability of such a large quantity of resources and services has resulted in the development of several novel applications, such as virtual reality [3], smart grids [4], and smart environments [5]. For delay-sensitive applications, however, the excitement quickly turns into stress as they try to satisfy the delay specifications. The issue becomes increasingly apparent and pressing when more and more intelligent machines and things get integrated into everyday human life, such as in the case of smart cities [6] and the Internet of Things [7]. The low-latency, location-aware, and mobile-support needs [8] cannot be met by the current cloud computing architecture [9]. Experts in the industry came up with the term "mobile edge computing" (MEC) to characterize the process of leveraging edge network capabilities to bring cloud-based services and resources closer to the end user, which is a potential solution to the problem. To meet these demands, mobile operators want to integrate the base station's processing, networking, and

G. B. Mohammed
Department of Computer Science and Engineering, Vardhaman College of Engineering, Hyderabad, India
e-mail: gousebaig@vardhaman.org

S. Shitharth (✉)
Department of Computer Science and Engineering, Kebri Dehar University, Kebri Dehar, Ethiopia
e-mail: shitharths@kdu.edu.et

G. Sucharitha
Electronics & Communication Engineering, Institute of Aeronautical Engineering, Hyderabad, India

G. Srivastava et al. (eds.), *Security and Risk Analysis for Intelligent Edge Computing*, Advances in Information Security 103,
https://doi.org/10.1007/978-3-031-28150-1_8

storage capabilities into a MEC platform. Like Cloudlet [10], MEC is an addition to, rather than a replacement for, the cloud computing concept. The MEC server can run the time-sensitive parts of the Programme, while the cloud server can handle the computationally heavy parts that can tolerate some delay. The promise of mobile edge computing (MEC) is that it will enable the billions of connected mobile devices to do computationally heavy tasks in real time at the network's edge. The characteristics that set mobile edge computing (MEC) apart are its close proximity to end users, its ability to enable mobility, and the widespread distribution of MEC servers around the globe [11]. Despite the many benefits, making MEC a reality is difficult because of administrative restrictions and security worries. The vision of MEC can't be realized without first exploring the essential conditions necessary for its realization and the opportunities that may present themselves in light of those conditions.

In contrast to cloud computing, edge computing sometimes lacks a centralised security mechanism, and when the number of terminal devices proliferates, the edge network as a whole may be vulnerable to malicious viruses owing to the insufficient protection of some of its individual nodes. If a computing task or service request is initiated in a MEC network, which is made up of many mobile devices with different identities, performances, and behaviours, the misbehaving nodes may act in selfish or malevolent ways, such as by rejecting relaying data packets. For example, a misbehaving node may drop a data packet while it is being relayed. When more and more bad actors join a MEC network, the entire system becomes less reliable. A 'Mirai' virus was used by hackers in 2017 to cause a widespread outage in the United States. The virus worked by infecting vulnerable devices and then used them to launch attacks on other network nodes. That's why fixing the security problems is the first step to making MEC work smoothly [12].

We set out to solve the existing security issue in MEC networks by creating a reliable trust assessment system that analyses users' and resources' identities, capabilities, and behaviors from a trust perspective. The success of a trust evaluation system determines not only whether or not MEC-related applications can be introduced to the market, but also whether or not tasks can be offloaded, resources can be scheduled, and shared, and vice versa [13]. The cornerstone of every thorough trust evaluation system is a reliable system for managing and analyzing reputation data and trust evaluation models. Trustworthy reputation manifestation, accurate reputation updating, and efficient reputation utilization are all possible with a well-planned reputation management system in a MEC network. Network operators are able to tell the difference between well-behaving de-vices and misbehaving de-vices, warding off attacks and facilitating the realization of a secure, reliable, and efficient edge computing network, thanks to a well-adapted universal trust evaluation model that increases the dependability of MEC network services and realizes the trust management of network nodes, etc. Due to a lack of research, many problems about mobile edge computing trust remain unresolved. Instead, much research has focused on cloud computing, specifically on how well users can trust one another in the cloud.

In spite of its many benefits, edge computing is not without its challenges. Some of them are data security and privacy [14], trust, compute at the edge node, offloading and partitioning, service quality, deployment methodologies, work-load, and rules. A device's ultimate rating, called "trust," is calculated using its peers' evaluations. It's possible to think of trust as the group's overall impression of a technology after considering everyone's experiences with it. The goals of this study are to (1) draw attention to the importance of trust management in Edge Computing architecture; (2) examine the state of the art in terms of trust management for edge computing; and (3) suggest a trust management system and compare its efficacy to that of current schemes. With the proliferation of Internet of Things (IoT) gadgets, we must inevitably adopt edge technologies [15]. To reduce latency, smart devices in edge computing share data with one another, but in IoT, the Cloud is responsible for ensuring the security of connected gadgets [16]. Because of this, a trust management system is needed to ensure the reliability of both the device and the data it provides and to identify the transmission of malicious data or information by malicious devices.

1.1 Contributions

The following contributions can be made as a result of carrying out this research:

- Highlighting trust management concerns in edge computing architecture and doing a complete analysis of existing systems and models that have already been presented to solve trust management difficulties.
- A unique trust management system is proposed to guarantee data security and rely heavily on edge devices for quick and efficient communication and processing.
- The suggested trust management system's implementation and assessment, along with a comparison to current systems, will be examined.

The remainder of the paper is structured as follows: Sect. 2 Literature reviews and a comparative study of current trust models. The "proposed trust management model" details the design and operation of the proposed framework and provides an explanation of the suggested approach in Sect. 3. Section 4 gives details of "Implementation and Results" that were achieved using the model. Section 6 wraps off with a "Conclusion" section.

2 Literature Survey

Mobile devices (smart phones, tablet PCs, etc.) have emerged as a vital resource in recent years, used for everything from education, entertainment, social networking, news and commercial payments [17]. Network video providers and wireless network operators face a significant challenge from this emerging trend since mobile devices have less capabilities than desktop computers, preventing customers from experiencing the same level of delight [18]. Mobile health care, mobile education, and mobile entertainment are just a few of the new service categories that have emerged in tandem with the rise of cloud computing. To provide these cloud computing services, data must be sent between mobile devices and the cloud processing centre. This throws extra strain on current networks and generates wholly new requirements for network bandwidth. The need for bandwidth is predicted to increase annually. In response to the increasing demands for high network load, high bandwidth, and low latency caused by the proliferation of mobile devices and the IoT, researchers have suggested the notion of mobile edge computing (MEC). They are also digging further into the MEC reference technology, module structure, and key application cases.

Most wireless edge devices have lesser capacity CPUs and memory, yet the new Internet links devices with limited resources. Wireless edge networks face this challenge every time they need to get data from a distant cloud service centre, hence several solutions have been developed and implemented. These include nomadic decentralised services [19], distribution calculations [20], and others. As a result, less processing time and power consumption on the part of the wireless network is required.

However, the data exchange between the service hub and the edge devices may experience increased delays due to the distribution computation. To solve this issue [21], suggest a distribution-based cloud-computing technique that makes use of adjacent nodes to finish allocating resources. Some data services in the wireless edge network can be efficiently improved by moving them to the terminal base station server, which also has the added benefit of reducing the network load, but this comes with its own set of security concerns. How to earn people' confidence while securely storing their sensitive information has emerged as a major challenge. With the widespread use of next-generation IT like cloud computing, big data, and the Internet of Things, encryption alone isn't enough to guarantee the safety of the wireless edge network. To overcome this, the wireless industry has begun incorporating reputation systems into their security protocols. Each node in an edge network often has a trust module for monitoring and assessing the behavior of its neighbors in a reputation system. As a result, the overall wireless edge network benefits from increased security, fault tolerance, and authenticity as malicious behavior are identified faster and with more precision. To prepare for the future, it is essential to build a data service security mechanism that makes use of the wireless edge network. Replay attacks, tampered messages, and forgery are only some of the security concerns that prevent users from relying on edge nodes for computation [22]. Thus, many academics work to design trust models to keep edge computing reliable. Author in [23] created a trust architecture for edge computing to enable large-scale distributed computing. Their multi-source feed-back approach incorporates the notion of global trust degree (GTD) by introducing three new core layers the network layer, the broker layer, and the device layer based on the foundations of direct trust and feedback trust from brokers and edge nodes. The term "multisource" refers to the fact that feedback was gathered from several sources, including both edge devices and service brokers. The effectiveness was measured using the global convergence time (GCT). We used the NetLogo event simulator and a custom similarity metric to conduct our experiments (PSM). Dependability was measured by calculating the task failure rate (TFR). Measurement theory is used in real time to assist in resource setup, and author in [24] presented a trust model to assess applications and a network node. To assess reliability with measurement error, confidence was introduced, and trustworthiness was defined to quantify probability quality. Multiple types of trust, including device trust, task trust, and inter-device trust, were considered when designing the framework. Additionally, a trust evaluation algorithm was devised to analyses the model, as well as a dynamic method of allocating resources for a job based on a trust threshold value while minimizing unnecessary duplication of effort. No updates on findings or implementation were provided. For the client/server architecture CuCloud (Volunteer Computing as a Service (VCaaS) System) that comprised both volunteer workstations and dedicated servers [25], introduced a model named SaRa (A Stochastic Model to Estimate Reliability of Edge Re-sources in Volunteer Cloud). The model was probabilistic in nature, making inferences about the reliability of individual nodes based on data gleaned by analyzing their past behavior. Task behavior and characteristics, such as success, failure, priority, etc., were among the most important parameters. Even though the model's probabilities were distributed at random, this strategy was verified using Google clusters and a testing environment with hundreds of computers. SaRa improved accuracy over other probabilistic models [26].

While cloud computing and peer-to-peer computing have received a great deal of attention and research [27], the field of edge computing networks has seen relatively few works on trust management mechanism. Instead, most research is scattered across other areas, such as mobile cloud computing, vehicle net-working, medical networking, etc. [28]. The concept of trust computability was initially articulated in the context of computer networks by [29]. Then, in 1994, Beth [30] postulated that trustworthiness may be evaluated through participation. Before any subsequent work could be done, Blaze [31] had to come up with the idea of "Trust Management" and build a trust management system around it. In terms of current research [32], concept integrates global and local reputations to create a trusted P2P network. They approach other nodes for their opinions on whether or not an entity may be trusted (including satisfaction with the transaction number of transactions, reliability of feedback, etc.). In addition, they highlighted the significance of node fraud detection and proposed that it be penalised, however they did not specify the penalty mechanism in the model itself. Kamvar's suggested global trust model Eigen Rep [33] uses a node's interaction history to determine its local trust value; from there, it iteratively calculates the trust degree between the adjacent nodes to determine the node's global credibility. Despite its ability to capture node behaviour, this model has some serious flaws, including its high complexity and low risk resistance. An entity can learn about the reputation of a target node in the P2Prep [34] trust model by gathering feedback from the target node's neighbours. The neighbouring node uses a voting method to choose the target node, and the entity chooses a node with a better reputation value to carry out the transaction. However, there are some doubts about the model's veracity on both sides. In addition to the impracticality of a centralised CA managing a database of all users and a future with billions of automobiles, centralised reputation management is another issue. For their proposed distributed reputation management system (DREAMS) [35], suggest using the vehicle's edge server as a means for the local administrator (LA) to carry out computing job requests made by the vehicle. You can find related work in [36], however in practice, you might not be able to trust these servers because the article doesn't account for the servers' track record of reliability. W. Data trust and node trust, developed by [37], are used to evaluate traffic data and mobile vehicle trustworthiness, respectively, as part of an anti-attack trust management system. However, the target vehicle gathers and disseminates information about its reputation. Similar in [38], was proposed that each vehicle works in tandem with its neighbors to carry out behavior detection, reputation calculation, and reputation performance. Segments of reputation are shared between vehicles. However, given the ever-changing context of a moving vehicle, reputational domains are more likely to be lost, shifted, or even invented in this way. It will be exceedingly challenging for a single vehicle with limited processing capabilities to cope with the challenge when the number of vehicles increases and the reputation computation effort grows from the collecting of reputation fragments to aggregation. In addition, there is a wealth of literature on establishing credibility by means of one's reputation (e.g [39]) or by means of the signing of a contract (e.g [40]), but the issues explored there are distinct from our own. To make better use of cloud resources, studies like [41] look at the trust assessment between users and cloud service providers.

3   RTEM System Architecture

In order to improve the efficiency and security of interactions in edge networks, a Reputation-based Trust Evaluation and Management (RTEM) system is presented. As shown in Fig. 1, the RTEM system architecture consists of edge servers, a cloud server, and mobile edge devices. The RTEM process is described in full below. When first joining the system, an edge device (whether fixed or mobile) will connect to the edge server that is geographically nearest to it. The edge server checks if the device is already in the system's edge device list (due to previous interactions, for example), and if so, it activates the device's status and reputation information for the subsequent task interaction in the system. If not, the edge server sends a query request to the cloud server. If the cloud server remembers the mobile edge device's past transactions, this indicates that the device has already registered with

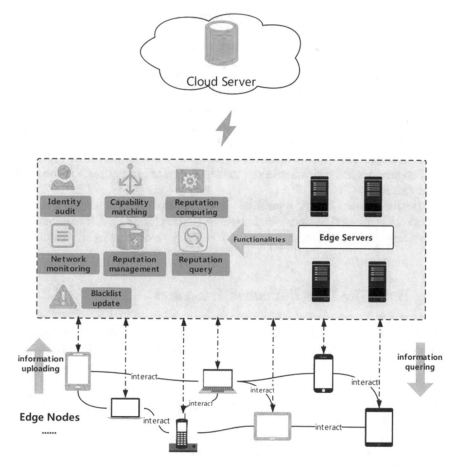

Fig. 1 Reputation-based Trust Evaluation and Management (RTEM) system

the RTEM system and generated the transaction record, prompting the edge server to download and store the generated reputation data. But if no devices have been added to the cloud, the edge server will do an audit of the device (using identity authentication and capability authentication) and, if successful, will give the device a preliminary reputation value. After the device has finished the task transaction and left the region for longer than the time threshold t, the edge server will upload its information record to the cloud server and trash it locally. More information regarding RTEM's primary features is provided below.

- In the context of reputation computing, a synthetic reputation segment for a mobile device is generated by collecting, weighting, and averaging several reputation segments within a certain time range (the so-called local reputation). For the most up-to-date estimate of one's global reputation, one must add the synthetic reputation segment to the most current estimate of one's local reputation.

- The most up-to-date reputation values are stored in a database for future use in reputation management. The associated record is synced with a cloud server and then removed locally if the mobile device leaves the region
- In order to prevent being compromised by rogue nodes, mobile devices can check their neighbors' reputation values to make sure they are communicating safely. For example, the system makes it easy to identify events like the current state of edge nodes and their interactions in the underlying planes thanks to the edge servers.
- Blacklist update: devices' reputation values are penalized or separated from the network if they fall below a certain level. Malicious gadgets are, if required, put to a regional blacklist that is broadcast to all nearby mobile gadgets.

3.1   Three-Tier Trust Evaluation Framework

In a three-stage trust assessment model, each stage is responsible for assessing a different facet of trust, trust in Identity, trust in Capabilities, and trust in Behavior. The levels of trust are depicted in a hierarchical form in Fig. 2. The following are definitions for the three distinct kinds of trust.

Identity trust (IT) Edge nodes are constantly changing and moving, making it hard to verify their true identities.

To ensure the authenticity and unambiguity of each node's identification in the network, the system requires real-name registration and delivers a voucher with a unique ID number to each device upon its initial login. The system will employ the identity authentication technology that has already been installed to determine if the operator's actual identification matches the digital identity when a device with a previously registered digital identity (i.e., ID number) reconnects to the network. If an edge device's authentication is successful, its identity trust (IT) is set to 1, and if not, it is set to 0. Devices that do not successfully authenticate through IT will be denied access to the network.

Capability trust (CT) Identifies the provider-supplied configuration data for the device's performance. Based on the configuration data, the system ascertains if the device's capabilities satisfy the service requester's task running criteria. In this study, we settled on the central processing unit (CPU), memory, disc space, and online time of a device at the network's periphery as the capability trust (CT) criteria for task matching against a reference service's demand. Parameter tables

Fig. 2 Evaluation of trust in
three stages

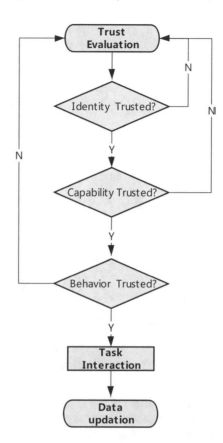

Table 1 Simulation parameters

Notation	Parameter	Value
α	Direct trust's weighting coefficient.	0.4
β	The coefficient of the trust's indirect value	0.6
δ	Coefficient of attenuation in time	1
ε and x	Objective penal factor coefficient	0.001
a	Penalty based on objective criteria	1.2
φ	Evaluation of the prescriptive interaction coefficient regulation	0.8
τ	The greatest permissible variation in grades	0.2

like Table 1 show common device capabilities and attributes. If the four capabilities
of the service provider match the task running requirements of the service requester,
then CT = 1. Otherwise, CT = 0. An overview of trust evaluation method is shown
in Fig. 3.

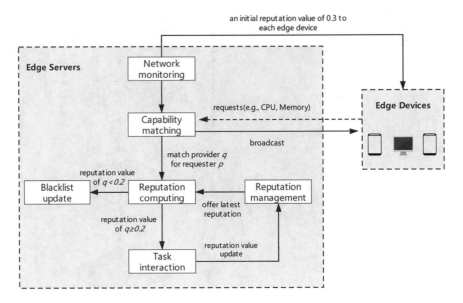

Fig. 3 illustrating an overview of the trust evaluation method.

4 Performance Evaluation

The effectiveness of the suggested trust evaluation model of edge computing networks is assessed here using the OMNeT++ platform. Memory (RAM): 8.00 GB, Central Processing Unit (CPU): Inter i5, Simulation Software: OMNeT++ 4.6. To get the optimal model, we conduct an experiment to determine the values of the parameters listed in Table 1. In an edge computing network, there are two types of nodes: those that are trustworthy and reliable, or honest, and those that are dishonest or unreliable, or malicious, and whose reputations suffer as a result. The service process of malicious edge node is frequently influenced by objective factors or even quits midway, damaging the security and stability of the system, leading to a low reputation value and successful transaction rate for the network as a whole. During the simulation, each experimental group completed fifty rounds of transactions using a network of one hundred identity-qualified edge nodes and one edge server S. The roles of the edges' nodes were picked at random (i.e., service provider or service requester). Requests for services fell into three categories: dumping work, downloading files, and computing activities.

5 Results and Discussion

The effectiveness of the suggested trust evaluation method is tested through experiments from three perspectives:

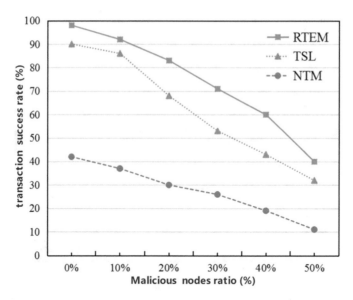

Fig. 4 Transaction success under multiple trust assessment mechanisms

Experiment 1 verification of the efficiency of our edge computing network's reputation-based trust evaluation mechanism (RTEM).

The success rates of transactions in edge networks employing our reputation-based trust evaluation mechanism (RTEM), a standard subject logic (TSL) trust evaluation mechanism, and no trust mechanism at all were compared (NTM). Figures 4 and 5 show how the transaction success rate varies with an increase in the ratio of malicious nodes and the ratio of oscillating nodes, respectively. If we compare the transaction success rate of an edge network with and without a trust assessment mechanism (NTM), we see that our RTEM-equipped edge network has a substantially higher success rate throughout a wide range of Malicious Node Ratio or Oscillation Node Ratio (see Figs. 4 and 5). On top of that, its findings are more accurate than those of the TSL trust evaluation method.

Without a trust mechanism, a network's transaction success rate is only approximately 41% and 68%, respectively, even if the ratio of malevolent nodes or oscillation nodes is zero. This is due to the absence of a capacity trust in the system, which would screen potential service providers (nodes) to determine whether or not they are qualified to take part in service transactions. However, in this scenario, the failure of a transaction is not due to any malevolent intent on the part of the edge nodes; rather, it is due to the incompatibility of the edge nodes' devices with the service type. The success rate of transactions, however, remains reasonably high thanks to the trust evaluation technique we suggest. Successful transaction rates of 40% can be maintained even when the fraction of malicious nodes exceeds 50%. TSL mechanism does better than NTM, yet we still perform better. As a result of the identical weight provided to all reputation segments in TSL's trust evaluation

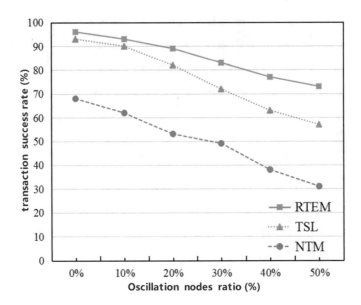

Fig. 5 Transaction success under multiple trust assessment mechanisms

procedure during reputation value updates, low-quality reputation segments might adversely affect reputation calculation and update, resulting in a lower rate of malicious device identification.

Experiment 2 examines the efficacy of both our local and international reputation models as a trust mechanism.

In order to maximise the usefulness of the data acquired from the reputation calculation experiment, we ran three separate sets of trials for a total of 50 iterations. In the first group of experiments, the edge nodes are the well-behaved, idealised nodes; in the second group, the edge nodes are the random, unpredictable nodes; and in the third group, the edge nodes are the oscillating nodes (i.e., Service quality is some-times good and sometimes poor). The local reputation values of trustworthy nodes converge over time, as shown in Fig. 6, whereas those of honest nodes, such as random nodes and oscillation nodes, slowly decrease as a result of their poor behavior. Our local reputation model was shown to be very sensitive to misbehaving nodes, demonstrating its usefulness in identifying and eliminating such behaviour in a network in a timely manner. Similarly, Fig. 7 shows that the global reputation value does not grow by rounds, even in the set of tests with honest nodes, reflecting the excellent limitations of time attenuation factor. Trust value from long-ago transactions, denoted by T (ti), accounts for only a small fraction of an organization's total global reputation value. In spite of this, our global reputation model shows that honest nodes have a far greater server trust than random and oscillation nodes, therefore the dishonest nodes are more likely to be detected.

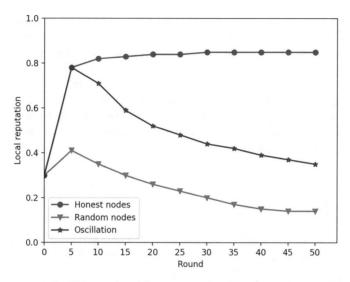

Fig. 6 Local reputation value of three different types of nodes over time is compared

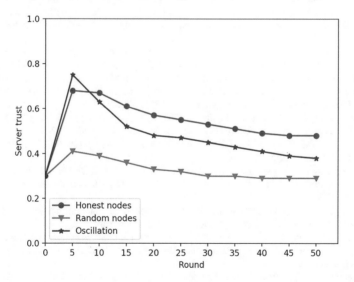

Fig. 7 Three types of nodes have been compared throughout time in terms of their global reputation value (server trust)

6   Conclusion and Future Scope

We propose a Reputation-based Trust Assessment and Management (RTEM) system for a reliable MEC network, wherein the edge server may keep track of, update, and provide the reputation values of mobility edge nodes in a given area. A three-tiered trust evaluation framework was presented to ensure that edge nodes entering the network system for service engagement are competent, qualified, and trustworthy. After that, we created a reputation-based trust rating model that took into account local and global contexts. We concluded by demonstrating the value of the reputation-based trust evaluation methodology we proposed.

The suggested process will be further examined for ways to increase its resilience against malicious assaults so that the trust evaluation mechanism can make more informed decisions. Additionally, we will focus on how the trust evaluation method may be used in a variety of contexts.

References

1. Mora-Gimeno FJ, Mora-Mora H, Marcos-Jorquera D, Volckaert B (2018) A secure multi-tier mobile edge computing model for data processing offloading based on degree of trust. Sensors 18:3211
2. Divakaran J, Prashanth SK, Mohammad GB, Shitharth D, Mohanty SN, Arvind C, Srihari K, Yasir Abdullah R, Sundramurthy VP et al (2022) Improved handover authentication in fifth-generation communication networks using fuzzy evolutionary optimisation with nano core elements in mobile healthcare applications. J Healthcare Eng, Hindawi. https://doi.org/10.1155/2022/2500377
3. Guo J, Li C, Luo Y (2020) Fast replica recovery and adaptive consistency preservation for edge cloud system. Soft Comput 24:14943–14964. https://doi.org/10.1007/s00500-020-04847-2
4. Yoosuf MS, Muralidharan C, Shitharth S, Alghamdi M, Maray M, Rabie OBJ (2022) FogDedupe: a fog-centric deduplication approach using multikey homomorphic encryption technique. J Sens, Hindawi. https://doi.org/10.1155/2022/6759875
5. Tabet K, Mokadem R, Laouar MR, Eom S (2017) Data replication in cloud systems. Int J Inf Syst Soc Change 8(3):17–33. https://doi.org/10.4018/IJISSC.2017070102
6. Jamali MAJ, Bahrami B, Heidari A, Allahverdizadeh P, Norouzi F (2020) IoT architecture BT. Towards Internet Things 21:9–31
7. Rani R, Kumar N, Khurana M, Kumar A, Barnawi A (2020) Storage as a service in fog computing: a systematic review. J Syst Archit 116:102033. https://doi.org/10.1016/j.sysarc.2021.102033
8. Fersi G (2021) Fog computing and Internet of Things in one building block: a survey and an overview of interacting technologies, vol 4. Springer, New York
9. Heidari A, Navimipour NJ (2021) A new SLA-aware method for discovering the cloud services using an improved nature-inspired optimization algorithm. PeerJ Comput Sci 7:1–21. https://doi.org/10.7717/PEERJ-CS.539
10. Shakarami A, Ghobaei-Arani M, Shahidinejad A, Masdari M, Shakarami H (2021) Data replication schemes in cloud computing: a survey. Springer, New York
11. Qin Y (2016) When things matter: a survey on data-centric Internet of Things. J Netw Comput Appl 64:137–153
12. Buyya R, Dastjerdi A (2016) Fog computing: helping the Internet of Things realize its potential. Computer (Long Beach, Calif) 49(8):112–116
13. Aberer K, Sathe S, Papaioannou TG, Jeung H (2013) A survey of model-based sensor data acquisition and management. In: Aggarwal CC (ed) Managing and mining sensor data. Springer, Boston
14. Shitharth S, Mohammad GB, Sangeetha K (2021) Predicting epidemic outbreaks using IoT, artificial intelligence and cloud. In: The fusion of Internet of Things, artificial intelligence, and cloud computing in health care, Internet of Things, vol 1, Issue 1. Springer, pp 197–222

15. Noel T, Karkouch A, Mousannif H, Al-Moatassime H (2016) Data quality in Internet of Things: a state-of-the-art survey. J Netw Comput Appl 73:57–81

16. Muralidharan C, Mohamed Sirajudeen Y, Shitharth S, Alhebaishi N, Mosli RH, Alhelou HH (2022) Three-phase service level agreements and trust management model for monitoring and managing the services by trusted cloud broker. IET Commun:1–12. https://doi.org/10.1049/cmu2.12484

17. Mazumdar S, Seybold D, Kritikos K, Verginadis Y (2019) A survey on data storage and placement methodologies for Cloud-Big Data ecosystem. J Big Data 6(1):15. https://doi.org/10.1186/s40537-019-0178-3

18. Sadri AA, Rahmani AM, Saberikamarposhti M, Hosseinzadeh M (2021) Fog data management: a vision, challenges, and future directions. J Netw Comput Appl 174:102882. https://doi.org/10.1016/j.jnca.2020.102882

19. Islam MSU, Kumar A, Hu Y-C (2021) Context-aware scheduling in fog computing: a survey, taxonomy, challenges and future directions. J Netw Comput Appl 180(1):103008. https://doi.org/10.1016/j.jnca.2021.103008

20. Bhageerath Chakravorthy G, Aditya Vardhan R, Karthik Shetty K, Mahesh K, Shitharth S (2021) Handling tactful data in cloud using PKG encryption technique. In: 4th smart city symposium, pp 338–343. https://doi.org/10.1049/icp.2022.0366

21. Hießl T, Hochreiner C, Schulte S (2019) Towards a framework for data stream processing in the fog. Inform Spektrum 42(4):256–265. https://doi.org/10.1007/s00287-019-01192-z

22. Naas MI, Lemarchand L, Raipin P, Boukhobza J (2021) IoT data replication and consistency management in fog computing. J Grid Comput 19(3):1–25. https://doi.org/10.1007/s10723-021-09571-1

23. Huang T, Lin W, Li Y, He LG, Peng SL (2019) A latency-aware multiple data replicas placement strategy for fog computing. J Signal Process Syst 91(10):1191–1204. https://doi.org/10.1007/s11265-019-1444-5

24. Li C, Tang J, Luo Y (2019) Scalable replica selection based on node service capability for improving data access performance in edge computing environment. J Supercomput 75(11):7209–7243

25. Aluvalu R, Uma Maheswari V, Chennam KK, Shitharth S (2021) Data security in cloud computing using Abe-based access control. In: Architectural wireless networks solutions and security issues. Lecture notes in network and systems, vol 196, Issue 1. Springer, pp 47–62. https://doi.org/10.1007/978-981-16-0386-0_4

26. Mohammad GB, Shitharth S (2021) Wireless sensor network and IoT based systems for healthcare application. Mater Today Proc, Elsevier. https://doi.org/10.1016/j.matpr.2020.11.801

27. Chen Y, Deng S, Ma H, Yin J (2020) Deploying data-intensive applications with multiple services components on edge. Mob Netw Appl 25(2):426–441. https://doi.org/10.1007/s11036-019-01245-3

28. Prasanth SK, Shitharth S, Praveen Kumar B, Subedha V, Sangeetha K (2022) Optimal feature selection based on evolutionary algorithm for intrusion detection. SN Comput Sci. https://doi.org/10.1007/s42979-022-01325-4

29. Li C, Wang YP, Tang H, Zhang Y, Xin Y, Luo Y (2019) Flexible replica placement for enhancing the availability in edge computing environment. Comput Commun 146:1–14. https://doi.org/10.1016/j.comcom.2019.07.013

30. Shao Y, Li C, Fu Z, Jia L, Luo Y (2019) Cost-effective replication management and scheduling in edge computing. J Netw Comput Appl 129:46–61. https://doi.org/10.1016/j.jnca.2019.01.001

31. Li C, Song M, Zhang M, Luo Y (2020) Effective replica management for improving reliability and availability in edge-cloud computing environment. J Parall Distrib Comput 143:107–128. https://doi.org/10.1016/j.jpdc.2020.04.012

32. Chennam KK, Aluvalu R, Shitharth S (2021) An authentication model with high security for cloud database. In: Architectural wireless networks solutions and security issues. Lecture notes in network and systems, vol 196, Issue 1. Springer, pp 13–26. https://doi.org/10.1007/978-981-16-0386-0_2

33. Mayer R, Gupta H, Saurez E, Ramachandran U (2018) FogStore: toward a distributed data store for fog computing. In: 2017 IEEE fog world congress, FWC 2017, pp 1–6. https://doi.org/10.1109/FWC.2017.8368524
34. Breitbach M, Schafer D, Edinger J, Becker C (2019) Contextaware data and task placement in edge computing environments. In: 2019 IEEE international conference on pervasive computing and communications (PerCom), March, pp 1–10. https://doi.org/10.1109/PERCOM.2019.8767386
35. Confais B, Parrein B, Lebre A (2018) A tree-based approach to locate object replicas in a fog storage infrastructure. In: 2018 IEEE global communications conference, pp 1–6. https://doi.org/10.1109/GLOCOM.2018.8647470
36. Lera I, Guerrero C, Juiz C (2018) Comparing centrality indices for network usage optimization of data placement policies in fog devices. In: 2018 3rd international conference on fog and mobile edge computing, FMEC 2018, pp 115–122. https://doi.org/10.1109/FMEC.2018.8364053
37. Confais B, Parrein B, Lebre A (2019) Data location management protocol for object stores in a fog computing infrastructure. IEEE Trans Netw Serv Manag. 16(4):1624–1637. https://doi.org/10.1109/TNSM.2019.2929823
38. Aral A, Ovatman T (2018) A decentralized replica placement algorithm for edge computing. IEEE Trans Netw Serv Manag 15(2):516–529. https://doi.org/10.1109/TNSM.2017.2788945
39. Hasenburg J, Grambow M, Bermbach D (2020) Towards a replication service for data-intensive fog applications. In: Proceedings of the 35th Annual ACM symposium on applied computing, pp 267–270. https://doi.org/10.1145/3341105.3374060
40. Guerrero C, Lera I, Juiz C (2018) Optimization policy for file replica placement in fog domains. Concurr Comput 9(1–20):2019. https://doi.org/10.1002/cpe.5343
41. Taghizadeh J, Ghobaei-Arani M, Shahidinejad A (2021) An efficient data replica placement mechanism using biogeographybased optimization technique in the fog computing environment. J Ambient Intell Humaniz Comput. https://doi.org/10.1007/s12652-021-03495-0

Network Security System in Mobile Edge Computing-to-IoMT Networks Using Distributed Approach

Eric Gyamfi, James Adu Ansere, and Mohsin Kamal

1 Introduction

The rapid growth in ubiquitous computing devices and human needs influence high demand for IoMT implementation across different disciplines and use cases. As predicted by CISCO [12], over 50 billion IoMT devices will be connected to the internet by the year 2020. The emergence of the Industry 4.0, domestic use, etc., of IoMT end-devices is forcing the rapid growth of the post-cloud era, which has intended increase data traffic at the network-ends. The critical and the need to process these data in a real-time nature is compelling most of the IoMT applications to be deployed at the edge of the network. However, the current IoMT end-devices face a server challenge of vulnerabilities to cyber-attacks [18] since they are mostly deployed over the public internet and hostile unsecured platforms. All these challenges exist in the IoMT due to resource-constraints such as limited storage, low processing power (CPU), and limited energy storage.

Most IoMT devices are also usually deployed in distributed environments which makes the end-nodes face many challenges from cyber and physical attacks. Due to these problems, secured measures are required to protect the IoMT devices from cyber-threats. IoMT security is prominent because data obtained from the sensors of

E. Gyamfi (✉)
University College Dublin, Dublin, Ireland
e-mail: eric.gyamfi@ucdconnect.ie

J. A. Ansere
Sunyani Technical University, Sunyani, Ghana
e-mail: jaansere@stu.edu.gh

M. Kamal
National University of Computer and Emerging Sciences, Islamabad, Pakistan
e-mail: mohsin.kamal@nu.edu.pk

© The Author(s), under exclusive license to Springer Nature Switzerland AG 2023
G. Srivastava et al. (eds.), *Security and Risk Analysis for Intelligent Edge Computing*, Advances in Information Security 103,
https://doi.org/10.1007/978-3-031-28150-1_9

these devices form the major driving force for critical decision making in both small and large scale industries. It is therefore important to investigate the high-security threat in IoMT networks and provide lasting solutions to damaging cyber-threats. IoMT end-devices do not have enough resources to comprehend with the modern trend of data manipulations and full implementation of requisite security systems. Since most of these devices are manufactured at a large scale with the same or similar architecture it becomes easy to replicate a successful attack on one IoMT end-device to another. Therefore, there is an urgent need for a security system that capitalise on Multi-Access Mobile Edge computing (MEC) technology to identify malicious attacks in the IoMT network and predict any possible future attacks. Edge computing platforms allow higher and complex algorithms to be performed at the network edge and bring computational intensive logic closer to the data source (IoT end-device). MEC address some of the problems associated with cloud computing, such as latency, bandwidth limit, location awareness and lack of mobility support problems encountered by cloud computing [4]. Due to the problems posed by cloud computing and the demand for real-time computational processing required by the IoMT end-devices, the MEC is much preferred over the cloud computing to support the security application developments. Figure 1 show the structure of the MEC platform.

Security systems are designed to protect the host devices from harm [26]. Both external and internal attacks are discovered in which external attacks are initiated outside the network, whereas, the internal attacks are launched through the compromised nodes that belong to the network. Security systems verify inbound

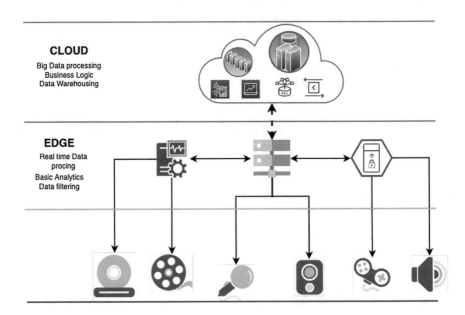

Fig. 1 Structure of the MEC platform

and outbound network connections to ensure whether or not they are intruders or legitimate connections. There are principally four general elements of any security system [25], which are,

- observation,
- analysis,
- detection, and
- protection

The patterns, network traffic and resources are monitored by the **observation** module. **Analysis** and **detection** could be the core part of security system that detects the attacks for the given algorithmic rule. The alarm module raised the alarm if an attacks are detected [22] and the **protection** model enforce the security configurations installed.

1.1 Security Problems in IoMT

The development of a massive pool of IoMT devices with different capabilities and constraints are the major influence of a high number of security threats [27]. Any security system designed for IoMT system is focused on two main approaches;

1.1.1 End-To-End Data Transmission Security

In IoT networks data transmissions, security is required to avoid interruption from unauthorized users. In general, the application of symmetric and asymmetric key-based cryptography mechanisms are used in data transmission, ensuring efficient security is a topmost priority for data transfer among IoT devices [8]. Several existing works offer a comprehensive study on the confidentiality and authenticity of data transmission to improve cryptosystem performances in IoT networks, such as flexible authentication, privacy-preservation and encryption algorithms that enhance efficient data transfer security [30]. Few research works seek to improve security in IoT networks, particularly on cybersecurity efficiency for IoT networks. Currently, extensive research is underway to implement lightweight encryption techniques to secure the end-to-end data transmission in IoT networks, ensuring high that IoTs are secured from cyber-physical threats [15]. Figure 2 shows the implementation of lightweight elliptic curve cryptography to protect end-to-end data transmission in IoMT utilising mobile edge computing [19].

1.1.2 Security for IoMT System Network

Aside from the use of encryption methods to secure the IoMT data transmission, the network and environment of the IoMT system must also be protected. Due

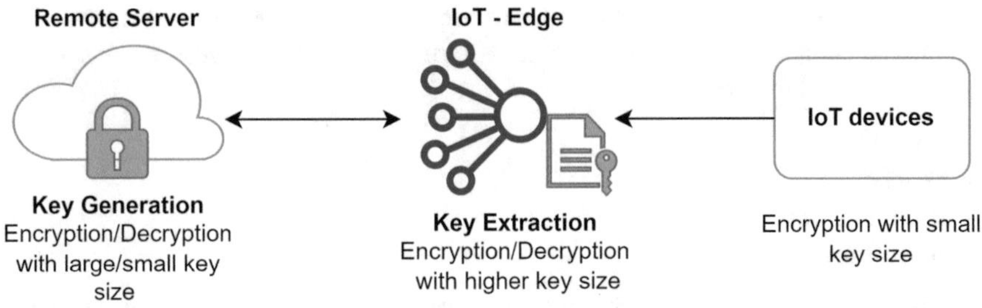

Fig. 2 ECC-based encryption for IoTs

to the resource constraints and the nature of IoMT System, Traditional network security systems cannot be implemented on IoMT. Different network attacks are also emerging due to the network structure of the IoMT system [36]. The next sections of this research review into details the various NIDS that can be implemented on the IoMT system to ensure maximum security of the IoMT system.

1.2 Threats in Communication and Protocols of IoMT

Different communication technologies have been developed to connect the IoMT and other networked devices to share information. These include Z-Wave, Near Field Communication (NFC), Internet Protocol Version 6 (IPv6), ZigBee, Bluetooth Low Energy (BLE), Low power Wireless Personal Area Networks (6LoWPAN), etc. They are short-range standard network protocols, while SigFox and Cellular are Low Power Wide Area Network (LPWAN) [3]. The ultimate aim of IoT is to connect all devices into one massive Internet network [44]. IoMT communication protocols are chosen based on resource constraints and capabilities. Processing capability, storage volume, short power life, and radio range are also some of the constraints mostly considered in choosing communication protocol for IoT. In selecting network protocol for IoMT network, criteria and benchmarks such as standards of the network, topology, Power consumption, range of coverage, and security mechanisms are considered. Most of the protocols above followed the IEEE 802.15.4a which was developed for a physical layer for sensor networks and similar IoT devices [43]. The principal goal for these new standards are energy-efficient data communications with data rates between 1 kbit/s and several Mbit/s; additionally, the capability for geolocation plays a significant role. All the aforementioned protocols are vulnerable to network attacks such as; Sniffing network traffic, Injection, Tampering/forging, Jamming, Exhaustion of battery, Collision, and Unfairness (link layer), Greed, homing, misdirection, black holes (network layer), and Flooding desynchronization (transport layer) [35].

1.3 Chapter Motivation

There are a number of research work published in diverse research areas on IoT security including security frameworks [5], security in context of e-Health [5], privacy issues [49], state of the art and security challenges [42], models, techniques, and tools, and attacks [11]. Some of these research works were published during the early stage of IoMT evolution. Most of the proposed techniques discussed in these papers have been implemented and commercialized. We also analyzed how the emergence of MEC has aided in the development of a sophisticated security system for the IoMT end-device and its network environment. This chapter address the following key challenges regarding the security of IoMT:

- This chapter provides an overview of some of the main IoMT security challenges.
- We analyze the possible security implementation strategies used in IoMT.
- This chapter also examine how the MEC can be used to protect the IoMT.
- Finally, we propose a security framework, provide an experimental demonstration, and benchmark the results.

1.4 Outline

The rest of this chapter is organized as follows: Sect. 2 contains the related works, Sect. 3 presents detailed information on multi-access mobile edge computing, its operation, the application in the context of IoMT. Section 4 describes proposed algorithms for attack detection and the implementation of a machine learning-based security system. The section also explains how it can fit into the IoMT end-device. Section 5 discusses the results obtained from the experimentation and performance evaluations. Section 6 concludes on the findings in the chapter.

2 Related Literature

2.1 Close Related Works

It is noted that, thousands of Zero-day attacks keep on evolving due to new protocols in IoMT. Over the years, different security approaches have been designed with different machine learning techniques. For Internet-of-Things (IoTM) networks, Khraisat et al. [17, 29] developed RDTIDS, a novel security system. The RDTIDS incorporates various classifier methods based on decision tree and rules-based principles, such as REP Tree, JRip algorithm, and Forest PA, based on decision tree and rules-based concepts. Diro et al. [13] used deep learning as a new way of detecting intrusion. They designed a distributed attack system which was proved by their experimental results to perform better than a centralised system. Multiple coordinated nodes were used for distributed approach and a single node was also used for the centralised method. The experimental results in their work show promise, but do not show clearly how the approach was implemented in an IoMT environment. In [45], a two-stage hidden Markov model was employed to group edge devices into four levels. A virtual honeypot device (VHD) was also used to store and organise log repository of all identified malicious devices that isolate the system from unknown attacks in future. This approach is similar to our method based on the multiple stages of cyber-attack detection and their test-bed used. Sudqi et al. [46] stated that, IDS is an integral part of any security for IoMT and Edge computing. In their paper, they presented a lightweight IDS based on vector space representation using multilayer perceptron (MLP). Their results show that using single hidden layer and a small number of nodes can help design an effective lightweight IDS for IoMTs using edge computing. [14] focused on creating a lightweight IoMT rules generation and execution framework. Their module consists of a machine learning rule discovery, a threat prediction model builder, and tools to ensure timely reaction to rule violation and standardized ongoing changes in IoMT Network traffic behaviour. They employed Random-Forest (RF) for the platform rule discovery and also perform a real time intrusion detection. Their application of lightweight technique is closer to our research, but did not demonstrate implementation of the Lightweight on a resource constrained IoMT device. Haddadpajouh et al. [21] used Recurrent Neural Network (RNN) deep learning to detect intrusion in their paper. Their research was based on evaluation of ARM-Based IoT device against malware. They evaluated their method with a Long Short Term Memory (LSTM) configuration with three different training sets. They, therefore, implemented intelligent security architecture using the Random Neural Network. Their security system was able to identify suspicious sensor nodes that were introduced into their IoMT network with an accuracy of 97.23%. The key evaluation their method was on the minimum impact of their system on IoMT's resources, power consumption, and performance.

2.2   State-of-the-Art Security for IoMT

The IoMT system has advanced in its security in communication for the past few years such that, the security framework cannot rely on traditional frameworks. Moreover, any application that targets the IoMT must consider the major constraints of the IoMT device. To secure the IoMT device, standardized security should be provided at different layers. At the link layer, the current IEEE 802.15 security protocol is the state-of-the-art security for IoMT systems [37]. The IEEE 802.15 was designed basically for defining the physical and the MAC sub-layer for low-rated personal area networks (usually used in IoMT communication). The IETF (Internet

Engineering Task Force) has been the force behind the development of IPv6 that works following the 6LoWPAN standards. 6LoWPAN bootstrap security, integrity, and privacy to preserve data exchange in the IoMT network. Low-power wide-area network (LPWA) which is designed to provide long-range wireless connectivity using the IPv6 technologies enhances security with symmetric key cryptography. This provides security for end-to-end device connectivity [40]. More resilient and vigorous standards are still underway to strengthen the security in IoMT communication networks. Therefore, researchers must consider the trends in security when designing security system for these devices that use such technologies. IP Security (IPsec) can also be used to secure the transport layer which consists of the HTTP, UDP, TCP-IP and CoAP protocols [1]. IPsec uses the Encapsulated Security Payload (ESP) protocol to enhance the confidentiality and integrity of IPv6 communication devices. ESP is embedded in the IP header of payloads in all IPv6 devices. IEEE 1905.1 is also a flexible and scalable Abstraction Layer for the various home networking technologies including power line communications, Wi-Fi, Ethernet, and MoCA 1.1 [38]. The Abstraction layer enhances control message exchange among IoMT devices operating with the IEEE 1901.1 standards. Despite all these standards that are currently been implemented, the IoMT devices still face cyber-threats from various attackers in diverse ways. This is because developers of IoMT applications are specific to the use-case of the implementation and does not pay more attention to the vulnerabilities in the IoMT's network environment.

3 Multi-Access Mobile Edge Computing (MEC) for IoMT

Multi-Access Edge Computing (MEC) brings the computational resources and IT services from a centralized cloud to the proximal edge of the network of the IoMT network. Instead of transferring the captured sensor data to the remote cloud servers for processing, the MEC analyzes, processes, and stores the data. Collecting and processing data closer to the IoMT end-device reduces latency and offers real-time performance.The preference for MEC over the traditional centralized cloud computing has risen dramatically in industries. The institutions such as the European Telecommunications Standards Institute (ETSI) and Open Edge Computing Initiative(OEC; openedgecomputing.org) [2, 16, 41] are creating standardization for the MEC. Cloud computing has a key drawback of low propagation delay [10]. Moreover, some other key characteristics that have given MEC attention are the context-awareness (the ability of the MEC to track the real-time information of the connected IoMT end-devices), energy-saving, and privacy/security enhancement (the ownership and management of users' data are separated in Cloud Computing, which causes the issues of private data leakage and loss) [47]. Figure 3 shows the structural layout of edge computing, Cloud or Remote server, and the IoMT. Some of the key application of MEC aside from security are;

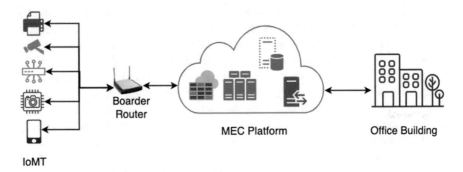

Fig. 3 IoMT to MEC connectivity

- Hosting of content, such as audio visuals
- Augmented reality
- Location network related services
- Data analysis and storage

MEC is executed based on a virtualized platform that employs network functionality virtualization (NFV), information-centric networks (ICN), and software-defined networks (SDN). NFV creates spontaneous virtual platforms to connect to connected IoMT end-devices [32]. In IoMT, computationally intensive tasks are transferred to the MEC server with the technique of *task offloading*.Since the MEC have enough resources, the computationally intensive algorithm of the security system are offloaded to the MEC. Offloading comes in two forms,*Partial Computational task offloading*, and *Binary task offloading*.

3.1 Distributed Security Strategies for IoMT Using MEC

The resource constraint nature of the IoMT makes it impossible to use an established security system to protect these cyber-physical devices [28]. Hence, the IoMT network require a custom Distributed Security systems (DSS) to protect them from attacks. DSS encompass multiple security software or devices installed on different nodes on the network(s) which, interact with each other to provide security. Most DSS have a cluster head that supervises the other DSS located on the IoMT nodes on the same network [7]. In a distributed IoMT MEC environment, co-operative intelligent systems are used to spread information across the network. Due to the resource constraints in IoMT, distributed data-flow programming models are used to build IoMT applications utilizing the edge computing platform or the fog computing platform [39]. IoMT systems permit distributed, autonomous decision making and intelligent data processing and analysis. However, security threats that are initiated to target directly or indirect against the MEC become detrimental to the whole network.

3.2 Lightweight Security System for IoMT

IoMT end-devices require a custom security system due to the discussed constraints. Lightweight security systems have been proposed by researchers as the best-practice to protect the IoMT end-device. According to Jan et al. [24], lightweight IDS does not mean simplicity, but the ability of the IoMT to perform its' designated duties without compromising the resource constraints. Masumi et al. [33] in his paper considered a lightweight IDS as a security system that is more flexible and powerful but small in size to enhance ease of implementation on the IoMT devices in a network. Lightweight DSS are created when the hosted IoMT end-device can fully perform the required security operations regardless of the resource constraints. Among some of the methods of creating lightweight NIDS are;

- Reducing the amount of operations needed by a standard security system by using new protocols.
- By optimizing a security system such that the scalar mathematical operations required are reduced.
- By making security systems more efficient so that fewer scalar mathematical operations are needed.
- Using contemporary algorithms that employ less computing power as opposed to traditional alternative ways.

4 Proposed Architecture and Methodology

In our proposed experimentation, a network intrusion detection system (NIDS) is created to represent our security system.The objective of this section is to implement security in an IoMT network and its environment utilising the MEC technology. Different machine learning methods can be adopted on both the MEC server and the IoMT device. The proposed algorithm uses One-Class support vector machine (OC-SVM) on the IoMT device and deep neural network (DNN) on the MEC server.The proposed design also follows the conceptual framework proposed in [20].

4.1 System Model and Design

Assuming $\mathcal{N}$ IoMT devices are erratically distributed where each has a security task to be executed. The set of IoMT devices is denoted as $\mathcal{N} = \{1, 2, \ldots, \mathcal{N}\}$. Let $\mathcal{M}$ denote the MEC server to offload the security tasks. The sets of MEC servers are represented as $\mathcal{M} = 1, 2, \ldots, \mathcal{M}$. Let α^i and $\alpha_{i,j}^{MEC}$ indicate the probable locations where the IoMT devices can offload the security tasks and MEC server location, respectively. Then, we have the following:

$$\alpha^i = \{0, 1\} \quad \forall i \in \mathcal{N}$$

$$\alpha_{i,j}^{MEC} = \{0, 1\} \quad \forall i \in \mathcal{N}, \forall j \in \{\mathcal{M}\} \tag{1}$$

$\alpha^i = 1$ indicates ith IoMT device executes the security task by itself, and $\alpha^i = 0$ otherwise, aMECij=1 denotes that the ith IoMT device offloads the security task to the jth MEC server (when $j \in \mathcal{M}$), and $\alpha_{i,j}^{MEC} = 0$ otherwise. Assuming each IoMT can execute its security task at most in one place, the we have

$$\alpha^i + \Sigma_{j \in \{\mathcal{M}\}} \alpha_{ij}^{MEC} = 1 \quad \forall i \in \mathcal{N}. \tag{2}$$

We can compute the ith IoMT device security task as β_i as:

$$\beta_i = \left[\delta_i, \gamma_i, \tau^{req}\right] \quad \forall i \in \mathcal{N} \tag{3}$$

Where $delta_i$ represents the CPU cycles of β_i to be computed, γ_i signifies the amount of data to transmit to the MEC server, and τ^{req} is the latency constraint. γ_i and δ_i are computed using $[\tau_{ij}^c = \delta_i / f_{ij}]$, where f_{ij} is the computational frequency of the MEC server. The offloading time is also computed as $[\tau_{ij}^r = \gamma_i / r_{ij}]$, where r_{ij} is the offloading data rate from i IoMT device to the MEC server. Based on the above analogy, we can formulate the total latency of the security offloading using the equation:

$$\tau_{ij}^c + \tau_{ij}^r = \frac{\gamma_i}{r_{ij}} + \frac{\delta_i}{f_{ij}} \leq \tau^{req} \tag{4}$$

The security task offloading latency constraint can be estimated using the equation:

$$\alpha^i \frac{\delta_i}{f_i^L} + \sum_{j \in \{\mathcal{M}\}} \alpha_{ij}^{MEC} \left(\frac{\gamma_i}{r_{ij}} + \frac{\delta_i}{f_{ij}}\right) \leq \tau^{req} \quad \forall i \in \mathcal{N}. \tag{5}$$

Where f_i^L is the IoMT devices' local computational frequency. The latency form the major part of the network between the IoMT and the MEC servers. However, the network transmission must be optimised to enhance the security task offloading in a real-time scenario.

4.1.1 Deep Neural Network

A machine learning approach known as **deep learning** employs an architecture with several hierarchical layers of non-linear processing stages. Depending on how the architecture is used, it may be divided into two types: deep discriminative

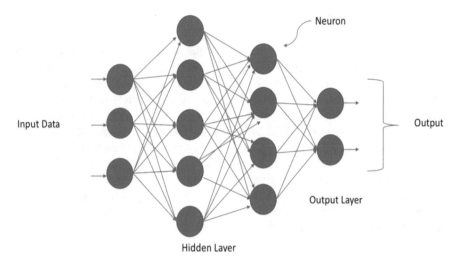

Fig. 4 Structure of DNN

architecture and deep generative architecture. With supervised learning, the deep discriminative architecture offers the same pattern categorization capabilities as traditional feed-forward artificial neural networks (ANN). Multiple hidden layers from the ANN structure can be added to the deep structure or deep neural network (DNN) as shown in Fig. 4. Due to the vanishing gradient issue, however, the augmented neural networks are ineffectively trained using back-propagation learning with a gradient descent optimization [6]. In our proposed model, we applied Long Short-Term Memory (LSTM) to design the neural network architecture, which is proposed by Hochreiter and Schmidhuber defined as: [48].

$$i_t = \sigma(W_{xi}x_t + W_{hi}h_{t-1} + W_{ci}c_{t-1} + b_i)$$
$$f_t = \sigma(W_{xf}x_t + W_{hf}h_{t-1} + W_{cf}c_{t-1} + b_f)$$
$$c_t = f_t c_{t-1} + i_c tanh(W_{xc}x_t + W_{hc}h_{t-1} + b_c) \qquad (6)$$
$$o_t = \sigma(W_{xo}x_t + W_{ho}h_{t-1} + W_{co}c_t + b_o)$$
$$h_t = o_t tanh(c_t)$$

σ is described as the logistic sigmoid function, whiles i,f,0 and c are the input gate, forget gate, output gate, and cell state respectively. W_{ci}, $W_c f$, and $W co$ represent weight matrices for peephole connections. In LSTM, three gates (i, f, 0) control the information flow. In DNN-LSTM model, the hidden layers are substituted by the LSTM Cells.

4.1.2 One-Class SVM

One-Class SVM is deployed to detect anomalies in a dataset. It works with implicitly mapping data vectors such as the input space in the feature space using a nonlinear kernel function. One-Class SVM constructs a nonlinear model using the data from the normal behaviour of a given system, such that data points that deviate from the normal model are identified as outliers. Wang et al. [31] created a hyperplane-based one-class SVM. In their work, image vectors used in the feature space were isolated from the origin by a hyperplane. The points that remained in the other half of the origin were identified as anomalies. Braun et. al., [23] also created One-Class SVM with hypersphere, whereby a minimum radius hypersphere is placed around the greater number of image vectors. The data vectors that were located outside of the hypersphere were considered as anomalies. The One-Class SVM algorithm can be mathematically formulated as

$$f(x) = \begin{cases} +1, & \text{if } x \in S \\ -1, & \text{if } x \in \overline{S} \end{cases} \tag{7}$$

Assuming a training dataset $(x_1, y_1), \ldots, (x_l, y_l) \in R^n \times \{\pm 1\}$, let $\phi : R^n \to H$ be a kernel map which transforms the training examples into the feature space H. Then, to separate the dataset from the origin, we need to solve the following quadratic programming problem:

$$min \left(\frac{1}{2} \|\Omega\|^2 + \frac{1}{vl} \sum_{i=1}^{t} \xi_i - \rho \right) \tag{8}$$

Subjected to:

$$y_i (\omega.\phi(x_i)) \geq \rho - \xi_i, \xi_i \geq 0, i = 1, \ldots, l \tag{9}$$

When this minimization problem is computed using Lagrange Multiplier, the decision function rule for a data point x produce the equation;

$$f(x) = sign((\omega.\phi(x)) - \rho) \tag{10}$$

The function $K(x, x_i) = \phi(x)^T \phi(x_i)$ is called kernel function. Since the results of the decision function only depend on the dot-product of the vectors in the feature space F (i.e. all the pairwise distances for the vectors). As long as a function K has the same results, it can be used instead. This is known as the kernel trick and it is what transform SVMs into such a great and powerful non-linear separable data points. Some of the most popular kernel functions used in SVM are linear,

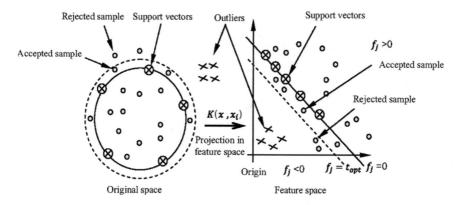

Fig. 5 One-class support vector machine with outlier

polynomial, sigmoidal, but the most commonly used is Gaussian Radial which can be represented by the equation:

$$K(x, x') = exp\left(\frac{-\|x - x'\|^2}{2\sigma^2}\right) \tag{11}$$

Fig. 5 demonstrates the use of One-Class SVM for classification and how the outliers are formed outside the accepted trained data.

4.1.3 Security Context-Awareness Creation with IoMT Neighborhood Discovery

In our experiment the MEC server is the head node of the network. In this sense, the MEC disseminate security context-awareness message to the connected IoMT nodes. The various IoMT nodes also sends a timely beacon message to tell the MEC that they are active in the network. This help MEC server to identify the IoMT end-devices that are under attac. The MEC also saves the information about the active neighbors on its list. The MEC takes advantage of beacons that are already exchanged by IoMT to obtain the contextual knowledge of the active nodes to avoid extra overhead. Assuming m denotes the MEC server, i represent all the connected IoMT end-devices, n is the total number of IoMT end-devices in the network, and a denotes the adjacency matrix, then the number of IoMT $G(m)$ is calculated as

$$G(m) = \sum_{i=1}^{n} a_{mi} \tag{12}$$

If a node is compromised due to attacks such that it is isolated from the network, it is assumed that it is a cut vertices $\phi(v)$ in the network. The cut vertices can be found using the Tarjan's algorithm [9] below.

$$\phi(v) = \begin{cases} 1 & \text{if } v = 1 \text{ is root and } N(v_m) > 1, \\ 1 & \text{if } v \text{ is not root with a neighbor } v \text{ such the } N(v_m)\backslash\{v\} \cap N(v_m)\backslash\{v\} \\ 0 & \text{otherwise} \end{cases}$$

$$(13)$$

4.2 Proposed Algorithm

To achieve this milestone, we executed a custom lightweight distributed OC-SVM on the IoMT devices. An LSTM machine learning based NIDS is configured to run on the MEC server. In our set-up, the Lightweight NIDS implemented on the various IoMT end-devices perform binary detection to check if the incoming network traffic contains intrusions. If the detected intrusion rate D_{IR} is greater than a preset threshold, the MEC is prompted to perform deep intrusion detection using the LSTM. If the MEC detects any attacks, they are compared to the results from the Lightweight NIDS on the various connected IoMT end-devices. A a security context-awareness is orchestrated to all the IoMT nodes on the server.

There are multiple sub-process such as $ScheduleRetrain()$ $Alert(Intrusion)$ $Context\text{-}AwarenessAlert(Intrusion)$, and $FeatureEng(RawData_{oc})$ that aid the execution of the DSS model. All this sub-processes operate on the MEC platform.

- The $Context\text{-}AwarenessAlert(Intrusion)$ sub-process sends security information to all the connected IoMT end-devices when the model detects attacks.
- The $ScheduleRetrain()$ also operates on the MEC. Its main duty is to retrain and deploy an updated model during off-peak hours with up-to-date datasets to keep the model from becoming obsolete.
- $FeatureEng(RawData_{oc})$ performs feature engineering on the captured network packets before use by the model.
- $Alert(Intrusion)$ sends attack flags to the context-awareness sub-process and the security logs for systems administrators.

Algorithm 1 demonstrate the coordination between the IoMT devices and the IoMT MEC platform.

Algorithm 1: The DSS function

---

1 $i_{min}, i_{max} \leftarrow Threshold_{range}$;
2 $RawData_{oc} \leftarrow Capture(NetworkTraffic)$;
3 $TestData_{oc} \leftarrow FeatureEng(RawData_{oc})$;
4 $TrainedModel_{oc}(TestData_{oc})$;
5 **Compute:** $Threshold(A_c) = \frac{Detected\ Anomalies}{Total\ TestSet} * 100\%$;
6 **if** $(A_c \geq i_{min}) \&\& (A_c \leq i_{max})$ **then**
7      $Send(RawData_{oc}, Rq_{oc}) \rightarrow MEC(RNN\ Model)$;
8      $A_c, Rp_{mec} \leftarrow Reponds(A_{cm}, Rp_{no})$;
9   **if** $(A_c \geq i_{min}) \& (A_c \leq i_{max}) \& (Rq_{oc} = Rp_{mec})$ **then**
10          $Context\text{-}Awareness\ Alert(Intrusion)$;
11          $ScheduleRetrain()$;
12          $Break$;
13   **else if** $(A_c > i_{max})$ **then**
14          $Alert(Intrusion)$;
15          $Break$;
16   **else**
17          $IGNORE\ TestData_{oc}$;
18 $return(A_c)$;

---

5 Experimental Result

5.1 Dataset Used

We used two different datasets to test our algorithm on two different stages. In the first stage, we employed a public dataset from the ADFA-NB15 [34] to train and tune our algorithms for higher accuracy and performance. In the second stage, we evaluate our algorithm base on live data from our testbed. ADFA-NB15, unlike the KDD-99, were generated based on current network threats. The ADFA-NB15 contains 175,341 training set and the testing set is 82,332 records from the different types of attacks and normal. The dataset contains 138 different protocols grouped under normal and abnormal. This demonstrates that the rapid growth in internet-enable device and technologies are enhancing different attacks with varieties of protocols and therefore security applications for IoTs must adapt to the skill of continuous and transfer learning. IDS for modern IoTs base on Machine Learning must adhere to dynamic Machine Learning and the dataset to frequent updates.

5.2 Evaluation Metrics

After the NIDS modelling, the final step was to evaluate the performance. The scientific way to benchmark the results of the model is by using a confusion matrix also known as the Contingency table. For a binary classification problem such as the OC-SVM consist of two rows and two columns as shown in Fig. 6. To better explain the Table, we can analysed it based on True Positive (TP), True negative (TN),

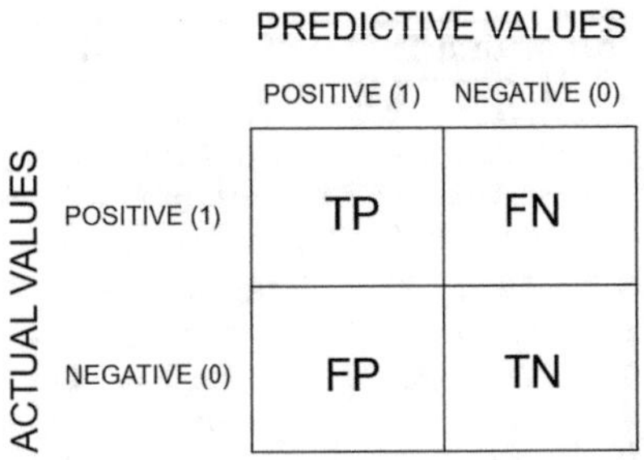

Fig. 6 Table of confusion
matrix

PREDICTIVE VALUES

POSITIVE (1)   NEGATIVE (0)

	POSITIVE (1)	NEGATIVE (0)
POSITIVE (1)	TP	FN
NEGATIVE (0)	FP	TN

ACTUAL VALUES

False Positive (FP) and False negatives. **True Positive (TP)** are predictions that were made by a model that represent the most accurate non-negative predictions that exist in a test set. Large TP indicate that a model is efficient and will be able to produce a good outcome when it is implemented. **True Negatives (TN)** represent the results of real negative outcomes that were predicted by a model compare to a test set. **False Positive (FP)** error happens when a model wrongly misinterprets a negative result as a positive outcome. It is sometimes called the False alarm. Whereas **False Negative (FN)** is when a model turns to reject a positive result as negative. The most detrimental parameters that affect the security model is FN and FP. It was deduced that, a large FN can render the security system of an IoT device prone to attacks. In the case of IoT security, False Positives are critical than False Negative.

The performances of the NIDS were evaluated based on the Detection Rate (DR) and the False Alarm Rate (FAR). The DR denotes the rate at which the NIDS detects intrusions. The FAR also signifies the ratio of misclassified instances of the model. As the value of DR increases, the FAR decreases.

$$DR(Precision) = \frac{TP}{TP + FN} \tag{14}$$

$$FAR = \frac{FP}{TN + FP} \tag{15}$$

The next parameter we considered in the performance evaluation of a model was the **Accuracy**. Accuracy measures the ratio of the correctly labeled subjects to the whole instances. It can be calculated as:

$$Accuracy(A_{cm}) = \frac{TP + TN}{TP + TN + FP + FN} \tag{16}$$

Using only the Accuracy to determine the performance of a security model may be Miss-leading. Therefore, parameters such as **Precision, Recall and F1 Score** were also considered. They represented mathematically as:

$$Recall/Sensitivity = \frac{TP}{TP + FN} \tag{17}$$

$$F1 Score = 2 \times \frac{Precision \times Recall}{Precision + Recall} \qquad (18)$$

To be more confident of our model and reduce the number of occurrence of offloading network traffic to the IoT-Edge for further detection, we use our computed precision ratio to determine the probability of FP in a real-time prediction.

$$Threshold = \frac{Detected Intrusion}{Total Test Set} * 100\% \qquad (19)$$

5.3 Results and Discussion

The data shown in Tables 1, 2, and 3 are the results from three different IoMT nodes connected to the MEC. The features of the dataset used in the OC-SVM model on the IoMT end-devices were reduced to 15 using principal component analysis to make the model lightweight. During the experiments the IoMT end-devices were represented with a raspberry pi zero. The MEC server was replaced with a windows core i7 device with 3.0 GHz four core processors. The MEC and the IoMT devices were connected together with a boarder router through a wireless connection.

Since the model on the MEC server perform deep intrusion detection, it was trained with all the features in the dataset. The results in Table 4 shows the performance of different learning rates of the RNN-LSTM.

Table 5 shows the various weighted averages obtained from different hidden layers. From Tables 4 and 5, we selected 0.01 as our learning rate and 90 as the size of our hidden layer because the high empirical values are achieved for precision, recall, F1-score and accuracy by keeping these parameters compared to the combination of others.

Table 1 Test result obtained by the IoMT node 1

Top 15 the features	DR	FAR	Accuracy
Decision tree	0.6626	0.2607	0.7990
KNN	0.5229	0.4175	0.6641
Naive Bayesian	0.2810	0.7862	0.3329
One-class SVM (rbf)	0.6857	0.0119	0.8281
One-class SVM (linear)	0.3147	0.1816	0.7362

Table 2 Test result obtained by the IoMT node 2

Top 15 the features	DR	FAR	Accuracy
Decision tree	0.6615	0.2516	0.7979
KNN	0.5229	0.4175	0.6641
Naive Bayesian	0.2810	0.7862	0.3327
One-class SVM (rbf)	0.7217	0.0126	0.8330
One-class SVM (linear)	0.3147	0.1816	0.7363

Table 3 Test result obtained by the IoMT node 3

Top 15 the features			
	DR	FAR	Accuracy
Decision tree	0.6640	0.2539	0.79687
KNN	0.5262	0.4144	0.6678
Naive Bayesian	0.2811	0.7855	0.3331
One-class SVM (rbf)	0.6961	0.0150	0.8324
One-class SVM (linear)	0.3147	0.1816	0.7363

Table 4 Preliminary test on the learning rate

Test on learning rate from 0.0001 to 0.1 (50 epochs)				
	0.0001	0.001	**0.01**	0.1
Precision (DR)	0.6	0.7	**0.74**	0.40
Recall	0.69	0.74	**0.77**	0.47
F1-score	0.65	0.70	**0.74**	0.39
Accuracy	0.6894	0.7362	**0.77**	0.47

Table 5 Effect of different hidden layers

Test on hidden layer from 10–90 (50 epochs)				
Size	Precision	Recall	F1-score	Accuracy
10	0.72	0.78	0.74	0.76
20	0.74	0.77	0.73	0.77
30	0.74	0.77	0.73	0.77
40	0.72	0.77	0.73	0.77
50	0.72	0.77	0.73	0.77
60	0.72	0.75	0.71	0.75
70	0.71	0.76	0.73	0.76
80	0.72	0.75	0.71	0.74
90	**0.80**	**0.80**	**0.77**	**0.80**

Table 6 Performance of RNN-LSTM on the MEC server

Test on the MEC server (500 epochs)				
Size	Accuracy	Precision	Recall	F1-score
77,302	0.83	0.83	0.83	0.81
64,419	0.83	0.83	0.83	0.81
38,651	0.83	0.83	0.83	0.81
25,768	0.83	0.83	0.83	0.81
12,884	0.84	0.83	0.84	0.82

5.4 Performance of RNN-LSTM NIDS on the MEC

Finally, we evaluated the performance of the RNN-LSTM on the MEC server. Different sizes instances were generated from the Test dataset. The performance of the MEC server is summarized in Table 6.

The weighted average of the results from the various sample test on the RNN-LSTM provided 83% Accuracy and a DR of 81%. This shows that out of every supply test record our model was able to detect 81% correctly.

6   Conclusion

Current and forthcoming IoMT systems need a reliable and efficient security with guaranteed quality of service (QoS) in practical scenarios. We have explored different security issues and implemented an NIDS models to demonstrate the feasibility of the proposed DSS in the IoMT system experimentally. However, designing a sophisticated DSS for resource-constrained IoMT systems whiles considering the critical metrics such as storage capacity, the energy required, and processing power needed to establish immense research area and remain an open challenge. Moreover, the experimental results have proved that a resilient DSS is achievable.

As cyber-attacks become more prevalent, IoMT application developers must conform to universal standards to build a security system for these cyber-physical devices. The idea of machine learning-based DSS and the MEC as a source of computing power for the IoMT should be considered a major component of IoMT security architecture. Future works will focus on adding online learning capabilities to the proposed model to keep it up-to-date, improve the it's performance, and optimize the execution time of the OC-SVM on the IoMT-end device.

References

1. rfc2401.txt.pdf. https://www.rfc-editor.org/pdfrfc/rfc2401.txt.pdf. Accessed 19 Feb 2020
2. Ai Y, Peng M, Zhang K (2018) Edge computing technologies for internet of things: a primer. Digital Commun Netw 4(2):77–86
3. Al-Sarawi S, Anbar M, Alieyan K, Alzubaidi M (2017) Internet of things (iot) communication protocols. In: 2017 8th international conference on information technology (ICIT). IEEE, pp 685–690
4. Alnoman A, Sharma SK, Ejaz W, Anpalagan A (2019) Emerging edge computing technologies for distributed iot systems. IEEE Netw 33(6):140–147
5. Ammar M, Russello G, Crispo B (2018) Internet of things: a survey on the security of iot frameworks. J Inform Secur Appl 38:8–27
6. Bibi M, Nawaz Y, Arif MS, Abbasi JN, Javed U, Nazeer A (2020) A finite difference method and effective modification of gradient descent optimization algorithm for mhd fluid flow over a linearly stretching surface. Comput Mater Continua 62(2):657–677
7. Bulaghi ZA, Navin AHZ, Hosseinzadeh M, Rezaee A (2020) Senet: a novel architecture for iot-based body sensor networks. Inform Med Unlocked 20:100365
8. Chanal PM, Kakkasageri MS (2020) Security and privacy in iot: a survey. Wirel Pers Commun 115(2):1667–1693
9. Chen R, Cohen C, Lévy J-J, Merz S, Théry L (2019) Formal Proofs of Tarjan's Strongly Connected Components Algorithm in Why3, Coq and Isabelle. In 10th International Conference on Interactive Theorem Proving, Portland, United States, pp. 1–19
10. Corcoran P, Datta SK (2016) Mobile-edge computing and the internet of things for consumers: extending cloud computing and services to the edge of the network. IEEE Consumer Electron Mag 5(4):73–74
11. Deogirikar J, Vidhate A (2017) Security attacks in iot: a survey. In: 2017 international conference on I-SMAC (IoT in social, mobile, analytics and cloud) (I-SMAC). IEEE, pp 32–37
12. Dilawar N, Rizwan M, Ahmad F, Akram S (2019) Blockchain: securing internet of medical things (iomt). Int J Adv Comput Sci Appl 10(1), 82–88

13. Diro AA, Chilamkurti N (2018) Distributed attack detection scheme using deep learning approach for internet of things. Futur Gener Comput Syst 82:761–768
14. Domb M, Bonchek-Dokow E, Leshem G (2017) Lightweight adaptive random-forest for iot rule generation and execution. J Inform Secur Appl 34:218–224
15. Dorri A, Kanhere SS, Jurdak R, Gauravaram P (2017) Blockchain for iot security and privacy: the case study of a smart home. In: 2017 IEEE international conference on pervasive computing and communications workshops (PerCom workshops). IEEE, pp 618–623
16. MECISG ETSI (2016) Mobile edge computing (mec): framework and reference architecture. ETSI, DGS MEC, 3
17. Ferrag MA, Maglaras L, Ahmim A, Derdour M, Janicke H (2020) Rdtids: rules and decision tree-based intrusion detection system for internet-of-things networks. Futur Internet 12(3):44
18. Gunduz MZ, Das R (2020) Cyber-security on smart grid: threats and potential solutions. Comput Netw 169:107094
19. Gyamfi E, Ansere JA, Xu L (2019) Ecc based lightweight cybersecurity solution for iot networks utilising multi-access mobile edge computing. In: 2019 fourth international conference on fog and mobile edge computing (FMEC). IEEE, pp 149–154
20. Gyamfi E, Jurcut A (2022) Intrusion detection in internet of things systems: a review on design approaches leveraging multi-access edge computing, machine learning, and datasets. Sensors 22(10):3744
21. HaddadPajouh H, Dehghantanha A, Khayami R, Raymond Choo K-K (2018) A deep recurrent neural network based approach for internet of things malware threat hunting. Futur Gener Comput Syst 85:88–96
22. Ioannou C, Vassiliou V (2020) Accurate detection of sinkhole attacks in iot networks using local agents. In: 2020 Mediterranean communication and computer networking conference (MedComNet). IEEE, pp 1–8
23. Jozani HJ, Thiel M, Abdel-Rahman EM, Richard K, Landmann T, Subramanian S, Hahn M (2022) Investigation of maize lethal necrosis (mln) severity and cropping systems mapping in agro-ecological maize systems in bomet, kenya utilizing rapideye and landsat-8 imagery. Geol Ecol Landscapes 6(2):125–140
24. Jan SU, Ahmed S, Shakhov V, Koo I (2019) Toward a lightweight intrusion detection system for the internet of things. IEEE Access 7:42450–42471
25. Jiang S, Zhao J, Xu X (2020) Slgbm: an intrusion detection mechanism for wireless sensor networks in smart environments. IEEE Access 8:169548–169558
26. Kamal M, Srivastava G, Tariq M (2020) Blockchain-based lightweight and secured v2v communication in the internet of vehicles. IEEE Trans Intell Transp Syst 22(7):3997–4004
27. Khan BUI, Olanrewaju RF, Anwar F, Mir RN, Najeeb AR (2019) A critical insight into the effectiveness of research methods evolved to secure iot ecosystem. Int J Inform Comput Secur 11(4–5):332–354
28. Khan ZA, Herrmann P (2017) A trust based distributed intrusion detection mechanism for internet of things. In: 2017 IEEE 31st international conference on advanced information networking and applications (AINA). IEEE, pp 1169–1176
29. Khraisat A, Gondal I, Vamplew P, Kamruzzaman J, Alazab A (2019) A novel ensemble of hybrid intrusion detection system for detecting internet of things attacks. Electronics 8(11):1210
30. Li M, Yu S, Zheng Y, Ren K, Lou W (2013) Scalable and secure sharing of personal health records in cloud computing using attribute-based encryption. IEEE Trans Parallel Distrib Syst 24(1):131–143
31. Mai W, Wu F, Li F, Luo W, Mai X (2021) A data mining system for potential customers based on one-class support vector machine. In: Journal of physics: conference series, vol 2031. IOP Publishing, p 012066
32. Mao Y, You C, Zhang J, Huang K, Letaief KB (2017) A survey on mobile edge computing: the communication perspective. IEEE Commun Surv Tutorials 19(4):2322–2358
33. Masumi K, Han C, Ban T, Takahashi T (2021) Towards efficient labeling of network incident datasets using tcpreplay and snort. In: Proceedings of the eleventh ACM conference on data and application security and privacy, pp 329–331

34. Moustafa N, Slay J (2015) Unsw-nb15: a comprehensive data set for network intrusion detection systems (unsw-nb15 network data set). In: 2015 military communications and information systems conference (MilCIS). IEEE, pp 1–6

35. Nawir M, Amir A, Yaakob N, Lynn OB (2016) Internet of things (iot): taxonomy of security attacks. In: 2016 3rd international conference on electronic design (ICED). IEEE, pp 321–326

36. Papaioannou M, Karageorgou M, Mantas G, Sucasas V, Essop I, Rodriguez J, Lymberopoulos D (2022) A survey on security threats and countermeasures in internet of medical things (iomt). Trans Emerg Telecommun Technol 33(6):e4049

37. Petrov V, Kurner T, Hosako I (2020) Ieee 802.15. 3d: First standardization efforts for sub-terahertz band communications toward 6g. IEEE Commun Mag 58(11):28–33

38. Poonia RC (2018) Internet of things (iot) security challenges. In: Handbook of e-business security. Auerbach Publications, pp 191–223

39. Premkumar S, Sigappi AN (2021) A survey of architecture, framework and algorithms for resource management in edge computing. EAI Endorsed Trans Energy Web 8(33):e15–e15

40. Raza U, Kulkarni P, Sooriyabandara M (2017) Low power wide area networks: an overview. IEEE Commun Surv Tutorials 19(2):855–873

41. Roman R, Lopez J, Mambo M (2018) Mobile edge computing, fog et al.: a survey and analysis of security threats and challenges. Futur Gener Comput Syst 78:680–698

42. Sain M, Kang YJ, Lee HJ (2017) Survey on security in internet of things: state of the art and challenges. In: 2017 19th international conference on advanced communication technology (ICACT). IEEE, pp 699–704

43. Sharma A, Sharma SK (2019) Spectral efficient pulse shape design for uwb communication with reduced ringing effect and performance evaluation for ieee 802.15. 4a channel. Wirel Netw 25(5):2723–2740

44. Siddiqui ST, Alam S, Ahmad R, Shuaib M (2020) Security threats, attacks, and possible countermeasures in internet of things. In: Advances in data and information sciences. Springer, pp 35–46

45. Sohal AS, Sandhu R, Sood SK, Chang V (2018) A cybersecurity framework to identify malicious edge device in fog computing and cloud-of-things environments. Comput Secur 74:340–354

46. Khater BS, Wahab AA, Idris M, Hussain MA, Ibrahim AA (2019) A lightweight perceptron-based intrusion detection system for fog computing. Appl Sci 9(1):178

47. Tabrizchi H, Rafsanjani MK (2020) A survey on security challenges in cloud computing: issues, threats, and solutions. J Supercomput 76(12):9493–9532

48. Werbos PJ et al (1990) Backpropagation through time: what it does and how to do it. Proc IEEE 78(10):1550–1560

49. Yang Y, Wu L, Yin G, Li L, Zhao H (2017) A survey on security and privacy issues in internet-of-things. IEEE Internet Things J 4(5):1250–1258

Wireless and Mobile Security in Edge Computing

Redowan Jakir Labib, Gautam Srivastava, and Jerry Chun-Wei Lin

1 Introduction

The number of wireless and mobile devices has been rapidly increasing for both commercial and personal use. Other usages for these devices are for improving city infrastructure, transportation arrangements, medical treatments, and many more. However, a high-performing computer platform is needed for these services, and due to cost issues and bandwidth shortages, an edge computing platform is preferred over cloud computing. Edge computing provides low-cost service, faster response time, improved insights, and better bandwidth availability to support emerging applications. Hence, edge computing is preferred over cloud computing due to these properties. According to a report in Statista in 2018, it was projected that in the United State of American (USA), the edge computing market size would increase from $84.3million to $1031 million in 2025 [1]. As for IoT devices, another report suggested that the full number of Internet of Things (IoT) devices increased from 11.2 billion in 2018 to 20 billion in 2020 [2].

In the case of emerging applications, edge computing gives a strong computing technology; however, its emergence is bringing more security threats every day. One of the most devastating cyberattacks that happened in recent years was the Mirai virus where around 65,000 edge devices were compromised in 2016 due to the exploitation of their weak authentication vulnerabilities [3]. These affected devices

R. J. Labib · G. Srivastava (✉)
Department of Mathematics and Computer Science, Brandon University, Brandon, MB, Canada
e-mail: LABIBRJ91@brandonu.ca; srivastavag@brandonu.ca

J. C.-W. Lin
Department of Computer Science, Electrical Engineering and Mathematical Sciences, Western Norway University of Applied Sciences, Bergen, Norway
e-mail: jerrylin@ieee.org

G. Srivastava et al. (eds.), *Security and Risk Analysis for Intelligent Edge Computing*, Advances in Information Security 103,
https://doi.org/10.1007/978-3-031-28150-1_10

were subsequently used to create botnets that launched DDoS attacks against edge servers, causing approximately 178,000 domains to go down [4]. These threats usually come from four angles.

1. Weak Computation Power: Edge devices possess weaker defence systems than computers, hence making them more vulnerable to cyber-attacks.
2. Attack Unawareness: It is quite difficult for users to realize when their edge devices get compromised, hence, most users may not be aware of an attack on an edge device.
3. OS and Protocol Heterogeneities: Since most edge devices use diverse operating systems and protocols with no established regulation, it is quite challenging to build a unified edge computing protection mechanism.
4. Coarse-Grained Access Control: There are four sorts of permissions in computer access control models: Read and Write (RW), Read Only (RO), as well as No Read and Write (NRW) [5].

   Due to more intricate systems and applications in edge computing, this approach would be difficult to implement, making it impossible to track who has access and who can access by doing what, when, and how.

The Mirai example as described demonstrates the huge security concerns in edge computing. This chapter discusses a detailed summary of security in edge computing [6], where we look at the different types of wireless attacks and security threats in smartphones and edge devices, as well as edge servers.

2 Types of Attacks and Defense Mechanisms

There are multiple types of threats and attacks that we require immediate attention of in-use systems. Figure 1 shows the major security threats and attacks over the last few years, where we see that Malware Injection and DDoS attacks take up 52% of the attacks, and Authorization attacks take up 22%. Man-in-the-Middle, Bad Data Injection, and Side-Channel attacks take up the rest 26% of the attacks.

These threats typically result from glitches, bugs, and design flaws in the software and firmware system. The main types of attacks are DDoS, Side-Channel, Malware Injection, as well as Authentication and Authorization attacks. Next, these various attacks are divided into sub-categories, as shown in Fig. 2.

2.1 DDoS Attacks and Defense Mechanisms

The well-known Distributed Denial of Services (DDoS) attack is a branch of attacks that can generate a traffic jam to prohibit legitimate users from accessing a service. This usually happens when an attacker delivers streams of packets to the victim's devices, which become too mainstream for the devices to manage. Once

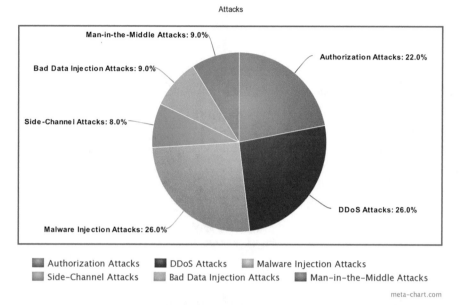

Fig. 1 Percentages of attacks that target wireless edge computing devices

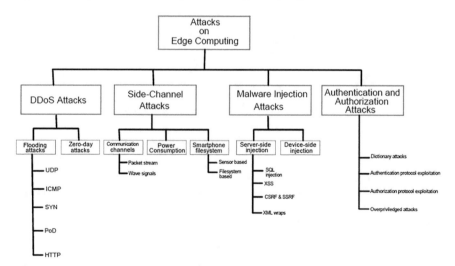

Fig. 2 Types of attacks on edge computing

the attackers have control of the compromised devices, they turn them into weapons to target the main edge servers, as shown in Fig. 3.

The Mirai virus is an example of a DDoS attack, where over 65,000 devices were compromised. Shortly after that, several other viruses were released. DDoS is the

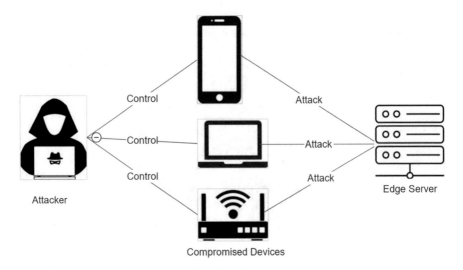

Fig. 3 Standard planning of a DDoS attack

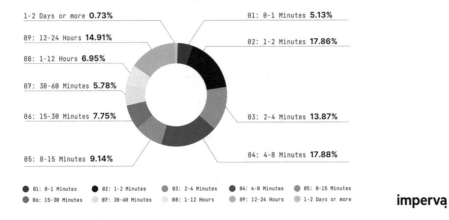

Fig. 4 Durations of DDoS attacks [7]

most common attack in the world, which usually lasts around 8 minutes on average, however, some DDoS may last 24 hours or more, as shown in Fig. 4.

DDoS attacks can be specified into two types.

(a) Flooding-based attacks: These are the most common type of DDoS attacks. Large amounts of flooding malicious network packets are transmitted to compromised devices in these types of attacks, which are mainly classified as:

- UDP flooding – A large number of UDP packets are sent to the edge server, causing the server to become unable to handle them promptly, which results in UDP service interruption [8].
- ICMP flooding – The ICMP protocol is used to create an attack by delivering a huge number of ICMP request packets to the server as quickly as possible with no wait time, causing the server to slow down completely [9].
- SYN flooding – A large number of SYN requests are sent to a server with a spoofed IP address, and the server waits for a confirmation from the spoofed IP address, which never arrives [10].
- PoD attack – An IP packet is produced with malicious content that is greater than the ordinary IP packet's maximum frame size and divides larger IP packets into numerous fragments which are sent to a server.

  The server must imitate all these fragments that eventually lead to the malicious IP packet, which consumes all the computation resources of the server, making it extremely slow [11].
- HTTP flooding – A server receives multiple HTTP GET, POST, PUT, or other requests. The server's computation power gets limited while managing a huge number of requests, hence, its services get throttled [12].

(b) Zero-day attacks: These are more advanced than flooding attacks and are difficult to implement. These usually occur when the server/device contains an error, bug, or flaw. This can result in memory corruption and the shutdown of the service. Per packet-based detection and statistics-based detection are the two types of defence solutions available against flooding-based attacks.

- Per-packet-based – Malicious packets transmitted by attackers are detected and filtered using per-packet detections. When a packet is discovered, the network has the option of dropping it before it reaches the intended server.
- Statistics-based – Attacks are detected using statistics-based detections when clusters of DDoS traffic appear. Since entropy computation requires a big quantity of DDoS traffic, these detections can only begin to work after a significant amount of DDoS traffic has already damaged the servers.
- To identify memory leaks in software, mechanisms such as pointer taintedness detection [13] and ECC-memory [14] are used as defence solutions against zero-day attacks. These mechanisms can provide security, but they cannot patch zero-day vulnerabilities, hence they can be ineffective against a vulnerability that is too difficult to exploit.

2.2   Side-Channel Attacks and Defense Mechanisms

Side-channel attacks infringe on a user's privacy by leveraging publicly available data. This occurs frequently because any public data can be linked to sensitive data, and this may be able to occur anywhere in any architecture that uses edge

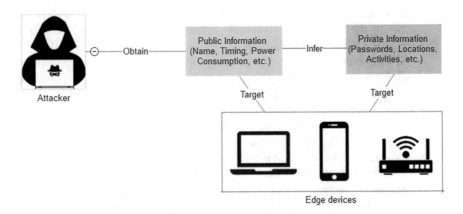

Fig. 5 Planning of a side-channel attack

computing. Figure 5 shows the planning of a side-channel attack, where the attack obtains public information through edge devices and infers private information.

The attacker gathers information regularly and feeds it to algorithms that produce the desired private data. Communication signals, power usage, and the smartphone filesystem are the most common side channels.

- Communication channels: Due to the extensive channel information, exploitation of communication signals has a significant probability of revealing the personal information of a user. The attacker continuously monitors the network traces and extracts the information. This can then be categorized into two types: the exploitation of wave signals as well as packet streams. The attacker can deduce personal data from a succession of packets since they carry wealth information. In a communication process, wave signals can reveal a user's information. Any given attacker can alter both or either of the output and/or input of an Internet of Things device from the physical layer via intentional electromagnetic interference (IEMI), bypassing typical integrity-checking techniques [15].
- Power Consumption: Power consumption is a measure of a system's power usage. Since different devices have distinct power consumption systems, this carries information about the device that consumes energy. Attackers can take advantage of this by using meters and oscilloscopes to acquire sensitive data. Smart meters can monitor a household's electric power consumption, and attackers can use the data from the meters to monitor activity in the household. An oscilloscope is used to measure the electronic information of a hardware device. Such devices have a secret key embedded in their chips, and if there is a flaw in the software system, attackers can exploit the system to acquire the key.
- Smartphone filesystem: Smartphones and mobiles are the key edge devices, making them more exposed to attackers. They use more modern operating systems and have access to more data. The /proc. filesystem and embedded

sensors in cell phones are both exploited by attackers. The kernel in Linux creates the /proc. filesystem, which includes information of the system for example interrupts and network data. Since it is a readable file, the access to the file system does not necessarily have the need of any specific permissions, leaving it vulnerable to side-channel attacks. Embedded sensors in smartphones perform a variety of jobs, but also raise security worries about sensitive data leakage. For example, the acoustic waves reflected by the fingertip recorded by microphones can be used by attackers to figure out the pattern lock [16].

Usually the attacks that are the most challenging are side channel to defend against among all sorts of attacks since they can be launched silently and passively. Although the main cause of the side-channel vulnerability is difficult to pinpoint, defences against side-channel attacks can be implemented in two ways: limiting access to side-channel data and protecting sensitive data from inference attacks.

- Data perturbation: K-anonymity perturbation algorithm is used for protecting private data from inference attacks. It modifies the identifier information of a piece of data before publishing its sensitive properties, making it different from other k−1 pieces of data, with all these k pieces of data forming an equivalence class [17].
- Limiting access to side channels: A method to protect against the software surface is side-channel obfuscation on the source code level. To mitigate side-channel attacks, researchers devised mitigation techniques based on TrustZone-enabled hardware, such as SGX. This approach primarily prevents unwanted access to TrustZone-protected side channels [18, 19].

2.3 Defence Mechanisms against Malware Injection Attacks

A malware injection attack is an action of effectively and secretly installing malware into a computing system. It is quite easy to bypass the firewall in edge devices and low-level edge servers, making this form of attack one of the most common and dangerous attacks. Figure 6 shows the usual planning of a malware injection attack, where the attacker injects malware into the connection between the servers and edge devices.

Server-side injections (attacks against servers) and device-side injections (attacks against devices) are the two types of malware injection attacks.

- Server-side injections: SQL injection, cross-site scripting (XSS), Cross-Site Request Forgery (CSRF), Server-Side Request Forgery (SSRF), and XML signature wrapping are the four types of server-side injection attacks.

  (a) SQL injection: This is a code injection method that causes the back-end databases to be destroyed. This is one of the most common methods for malware injection, in which attackers combine query strings with escape characters (such as quote marks) in restrictions. After the escape characters,

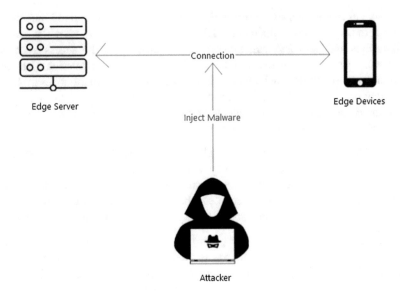

Fig. 6 Planning of a malware injection attack

the server may inadvertently execute anything the attacker types. Malicious scripts can also be injected through the SELECT... INTO OUTFILE command [20].

(b) XSS: This is a client-side attack in which the attacker injects harmful code into the data content using HTML/JavaScript codes that could be automatically executed by the server. When servers do not filter code from data content, this is what normally happens.

(c) CSRF & SSRF: CSRF is an attack in which a web application forces an edge server to do unauthorized actions. SSRF is an attack that uses edge servers to read or change internal resources. Attackers take advantage of the coarse-grained verification and pose as a "legal" server to send commands to other servers without being detected, exposing those systems to attacks. These attacks are mostly directed at internet infrastructure.

(d) XML: When an edge computing infrastructure adopts the Simple Object Access Protocol (SOAP) as its communication protocol, messages are sent in XML format. An attacker intercepts a valid XML message, produces a new tag, and inserts a duplicate of the original message into the new tag to construct a tag-value pair. The attacker then replaces the original values in the original message with malicious codes, then merges the changed original message with the tag-value pair by inserting the new pair before the original message's regular pair. When the server receives this message, it runs the malicious code that the attacker has injected [21].

- Device-side injections: Attackers have devised a variety of malware injection methods since IoT devices are highly sensitive in terms of hardware and firmware. Attackers can take advantage of zero-day vulnerabilities to obtain remote code execution (RCE) or command injection. Attackers typically hunt for a bug in a device's firmware mechanism, and once found, infect the firmware through networking. Smartphones are protected from malware injections because their operating systems, such as iOS and Android, have an app-isolation mechanism that ensures that each application is virtually isolated on memory and that no application can access the resources and contents of other applications unless the user grants permission at the kernel level [22]. Attackers have discovered a means to circumvent this safeguard by utilizing third-party malicious libraries, which are not only more potent but also less likely to be caught. These malicious libraries usually open a backdoor for code injection. This requires users to install these malicious applications for their devices to be attacked, however, there are suspicious websites available that can bypass the security of Android WebView and remotely inject malicious applications into an Android device [23].

Server-side injections mainly happen due to errors in the protocol-level designs, while device-side injections mainly happen due to errors in both code-level design and coarse-grained access control models. Defences against server-side injections: There are defence mechanisms for all four types of attacks.

(a) SQL injection: There are two mechanisms available to defend against SQL injection attacks. The first method is to install a proxy filter on PHP-based servers to prevent any unauthorized SQL queries from being executed [24]. The second method is to set up a proxy filter in servers built on Java [25].
(b) XSS: Manually setting hard-coded rules on the client side to prevent malicious codes from being executed and implementing ISR to render harmful codes harmless are two defence methods against XSS attacks [26].
(c) CSRF & SSRF: The referenced header-checking CSRF defence method mandates the client side to deliver an origin header to defend against the login CSRF [27]. The SSRF protection technique relies on clients' credentials being embedded in requests [28] and the static web application firewall (WAF) approach is enabled [29].
(d) XML: XML has a variety of protection methods. Wrapping attacks are mitigated by a W3C XML Schema [30]; a side-channel-based detection system that filters wrapping attacks by calculating the frequency of each node in a requested service [31]; and a technique based on positional tokens that identify incoming wrapping assaults [32].

Defences against device-side injections: The main threat for IoT devices comes from firmware modification attacks. The most recent and widely used defence technique for securing firmware is to use the memory protection unit (MPU) under RISC-V to segregate sensitive data and instructions from unrelated data [33]. Several mechanisms are available to identify malicious libraries and malware. A library detection mechanism has been implemented to detect security vulnerabilities and malicious behaviours in Android libraries [34]. Unfortunately, there are no modified defence mechanisms to protect against suspicious websites that can bypass the security of Android WebView.

2.4 Authentication and Authorization Attacks and Defense Mechanisms

Authentication is the process of confirming a user's identity to access services. Authorization is the process of granting users access to a content's access rights and privileges. For identity verification, authorization is frequently followed by authentication. The attacker bypasses authentication and permission systems in these types of attacks, obtaining unauthorized access to sensitive data, as shown in Fig. 7.

Authentication and authorization attacks can be categorized into four types:

- Dictionary attacks: This attack is the most common in authentication and authorization attacks which happens when the attacker employs a password dictionary to find a match and crack the password-enabled authentication system. It is relatively straightforward to retrieve passwords because dictionaries are available in open-source communities [35]. To figure out the passwords, attackers utilize Bluetooth's three-party password-authenticated key exchange (S-3PAKE) protocol [36, 37]. Biometric security systems can also be circumvented by using the covariance matrix adaption evolution strategy (CMA-ES), differential evolution (DE), and particle swarm optimization (PSO) to create a synthetic master print that could match the target's fingerprints [38].
- Authentication protocols exploitation: Attackers often target the vulnerability in the WPA enterprise authentication protocol and attack the WPA [39]. They also target the TLS authentication mechanism, which might lead to impersonation and

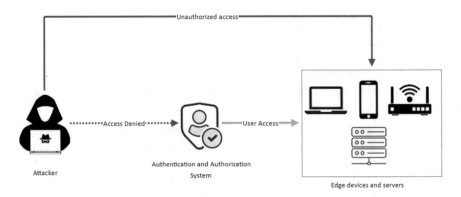

Fig. 7 Bypassing the authentication and authorization system

downgrade attacks in practice [40] In addition, attackers have been discovered to target OSes and platforms to compel nonce reuse in WPA2, allowing them to replay, decode, and fabricate authentication messages [41, 42]. Besides the WPA/WPA2 mechanism vulnerabilities, flaws in the authentication methods in 4G and 5G networks were also discovered. The 4G network in an edge device can be exploited by tampering with six network functional safety problems, denying the user 4G services [43]. By conducting a man-in-the-middle (MITM) attack and illegally accessing crucial unencrypted data, the security protocol in the 4G LTE network can also be exploited [44]. Attackers can also leverage the 4G LTE protocols' attach, detach, and paging methods to spoof a user's location [45]. A user's location can also be tracked, and fabricated paging messages can be injected into the edge device by exploiting the security protocol in 5G network implementations [30].

- Authorization protocols exploitation: The vulnerability of design weakness or logic flaws in authorization protocols is often exploited by attackers. A widely used authorization protocol called OAuth is designed for multiparty authorization [27, 28]. OAuth entails three parties: the user, the service provider, and the relying party. The goal of OAuth is to allow the service provider to access the user's resources (which are held in the relying party) only if the user gives the service provider access rights. Fixation attacks, which occur when the service provider requests token approval from the dependent party, can be used to exploit this authorization mechanism [46]. Attackers can simply acquire unauthorized access to the user's sensitive data.
- Overprivileged attacks: There have been multiple cases of overprivileged issues in authorization systems. Smart home platforms are vulnerable to these attacks, which allow attackers to gain unauthorized access and manage platform functions such as altering door pins or falsely activating a fire alarm, among other things [47], and also bypass the authorization system and control smart home devices [48].

The usage of weak passwords in authentication protocols is the primary cause of dictionary attacks, but protocol-level/implementation-level defects are the primary source of the other three attacks. Strengthening the password verification process or adding a stronger authentication layer are simple ways to defend against dictionary attacks, however, the defence mechanisms for the other three attacks require code-level analysis or patching of the security protocols.

- Dictionary attacks: Since these types of attacks are the most common, it is easy to defend against such attacks. These attacks typically fail most of the time and require lots of time to implement. The attacks can easily be avoided by putting a complicated password or by enabling two-factor authentication [49]. Most devices and services support two-factor authentication, which uses a variety of attributes as the second authenticator, such as fingerprints, a face scanner, or a code sent over SMS or email. All the second authorizations require human interactions; hence it is tough for automated edge devices to operate. However, it is not 100% guaranteed to be safe, attackers can still bypass the two-factor authorization protocol and gain unauthorized access to sensitive data [50, 51].

- Authentication protocols: The key defence methods are improving the security of communication protocols or securing cryptographic implementations. To prevent the decryption of WPA communication, an active jammer and wireless packet injection can be used [52]. The threat of various vulnerabilities in the WPA/WPA2 protocol is reduced by adapting public key cryptography [53]. TLS defence mechanisms have been developed to minimize the attacks [54]. To strengthen 4G and 5G network protocols, a three-step approach is available: The first step is to signal transfer by adding a slim layer; the second step is to decouple domains in the 4G radio resource control (RRC) layer, and the third step is to coordinate comparable functions across systems. Following this three-step solution protects a user from denial of service in 4G networks [43]. In 5G heterogeneous networks, SDN can be used for quick authentication [55]. The use of RF fingerprints as a piece of evidence to defend against impersonation-related attacks has resulted in the development of a lightweight, scalable, secure cross-layer authentication architecture for 5G [56].
- Authorization protocols: Most of the authorization security risks are found in the code-level system. To examine and fix OAuth implementation weaknesses, a static code analysis tool has been developed [57], and to prevent the current OAuth APIs from being exploited, an application-based OAuth Manager framework has been created [58].
- Overprivileged attacks: The majority of overprivileged concerns occur in edge systems and devices. The mismatches between an IoT application's implementation and its description are checked using an NLP-based approach to identify overprivileged attacks [59]. Another taint-based analysis has been developed to track and prevent any sensitive data leakage due to overprivileged designs of IoT applications [60]. To keep track of probable data leaks from overprivileged apps, an information tracking technique is used [61]. IoTGuard, a model-checking-based tool, is available to automatically identify overprivileged applications [62]. Another approach for preventing attacks is to compare the contexts of activities with previous ones to discover any suspicious inconsistencies [63]. The use of environmental situation oracles (ESOs) to enforce IoT access control in situational environments is also possible [63, 64]. Another typical method to defend against overprivileged attacks is to strengthen the current permission systems in mobile OSes. For Android, there is a semantic permission generator that understands an application's description and grants permissions to the application based on the interpretation [65], the application code that requests sensitive permissions is then placed in a system-level sandbox [66], which would then be isolated for further analysis to prevent any overprivileged permissions. Another mechanism called 6thSense is used to prevent overprivileged applications from exploiting Android sensors by applying three machine learning models, i.e., Naïve Bayes, logistic model tree (LMT), and Markov chain [67]. Another model is available that uses a graph abstraction algorithm and a logical reasoning algorithm to detect the overprivileged components of an Android application [68].

3 Conclusion

In this chapter, we looked at different types of attacks that usually happen in edge devices and servers, as well as the accompanying defence mechanisms. The defence mechanisms have been practiced and tested in every way possible by researchers, however, they are not 100% guaranteed to protect a user from being a victim of a cyber attack. As technology evolves every day, so does the security challenge of technological systems, researchers are still experimenting and testing new defence mechanisms to tackle today's security challenges.

References

1. Segment (in Million U.S. Dollars) (2017) Edge computing market size forecast in the United States from 2017 to 2025. Statista
2. Billion Units (2019) Number of IoT devices in use worldwide from 2009 to 2020. Statista
3. Antonakakis M et al (2017) Understanding the Mirai botnet. Proc 26th USENIX Secur Symp, Vancouver, BC, Canada
4. Financial impact of Mirai DDoS attack on dyn revelaed in new data (2017) [Online]. Available: https://www.corero.com/blog/financial-impact-of-mirai-ddos-attack-on-dyn-revealed-in-new-data/
5. Barth A, Jackson C, Mitchell AJC (2008) Robust defenses for cross-site request forgery. Proc 15th ACM Conf Comput Commun Secur (CCS):75–88
6. Xiao Y et al (2019) Edge computing security: state of the art and challenges. Proc IEEE 107(8):1608–1631
7. Venturebeat. Imperva. [Online]. Available: https://venturebeat.com/2021/09/04/the-median-duration-of-ddos-attacks-was-6-1-minutes-in-the-first-half-of-2021/
8. Xiaoming L, Sejdini V, Chowdhury AH (2010) Denial of service (DoS) attack with UDP flood. School Comput Sci Univ Windsor, Windsor, ON, Canada
9. Kolias C, Kambourakis G, Stavrou A, Voas AJ (2017) DDoS in the IoT: Mirai and other botnets. Computer 50(7):80–84
10. Bogdanoski M, Shuminoski T, Risteski AA (2013) Analysis of the SYN flood DoS attack. Int J Comput Netw Inf Secur 5:1–11
11. Elleithy K, Blagovic D, Cheng W, Sideleau AP (2006) Denial of service attack techniques: analysis, implementation and comparison. J Syst Inform 3:66–71
12. Dhanapal A, Nirthyanandam AP (2017) An effective mechanism to regenerate HTTP flooding DDoS attack using real time data set. ICICICT:570–575
13. Qin E, Lu S, Zhou AY (2005) SafeMem: exploiting ECC-memory for detecting memory leaks and memory corruption during production runs. Proc 11th Int Symp High-Perform Comput Archit:291–302
14. Chen S, Xu J, Nakka N, Kalbarczyk Z, Iyer ARK (2005) Defeating memory corruption attacks via pointer taintedness detection. Proc Int Conf Dependable Syst Netw (DSN):378–387
15. Selvaraj J, Dayanikh GY, Gaunkar NP, Ware D, Gerdes RM, Mina AM (2018) Electromagnetic induction attacks against embedded systems. Proc Asia Conf Comput Commun Secur:499–510
16. Zhou M et al (2018) PatternListener: cracking Android pattern lock using acoustic signals. Proc ACMSIGSAC Conf Comput Commun Secur (CCS):1775–1787
17. Sweeney L (2002) K-anonymity: a model for protecting privacy. Int J Uncertainty Fuzziness Knowledge Based Sys 10(5):557–570
18. Piessens RSAF (2017) The Heisenberg defense: Proactively defending SGX enclaves against page-table-based side-channnel attacks. [Online]. Available: https://arxiv.org/abs/1712.08519

19. Fu Y, Bauman E, Quinonez R, Lin AZ (2017) Sgx-Lapd: Thwarting controlled side channel attacks via enclave verifiable page faults. Proc Int Symp Res Attacks Intrusions Defenses:357–380

20. Anley C (2002) Advanced SQL injection in SQL server applications. Proc CGISecurity:1–25

21. McIntosh M, Austel AP (2005) XML signature element wrapping attacks and countermeasures. Proc Workshop Secure Web Services:20–27

22. Gallmeister B (1995) POSIX.4 programmers guide: programming for the real world. O'Reilly Media, Inc, Sebastopol

23. Li T (2017) Unleashing the walking dead: understanding cross-app remote infections on mobile WebViews. Proc ACM SIGSAC Conf Comput Commun Secur (CCS), New York:829–844

24. Halfond WJ, Viegas J, Orso A (2006) A classification of SQL injection attacks and countermeasures. Tech Rep

25. Livshits VB, Lam MS (2005) Finding security vulnerabilities in java applications with static analysis. Proc USENIX Secur Symp 14:18

26. Gupta SGAB (2017) Cross-SITE Scirpting (XSS) attacks and defense mechanisms: classification and state-of-the-art. Int J Syst Assur Eng Manag 8(1):512–530

27. Hammar-Lahav E (2010) The OAuth 1.0 protocol. Document RFC 5849, IETF

28. Hardt D (2012) The OAuth 2.0 authorization framework. Document RFC 6749, IETF

29. Srokosz M, Rusinek D, Ksiezopolski B (2018) A new WAF-based architecture for protecting Web applications against CSRF attacks in malicious environment. Proc Federated Conf Comput Sci Inf Syst (FedCSIS):391–395

30. Hussain SR, Echeverria M, Chowdhury O, Li N, Bertino E (2019) Privacy attacks to the 4G and 5G cellular paging protocols using side channel information. Proc NDSS:1–15

31. Jensen M, Meyer C, Somorovsky J, Schwenk J (2011) On the effectiveness of XML Schema validation for countering XML Signature Wrapping attacks. Proc 1st Int Workshop Securing Services Cloud (IWSSC):7–13

32. Kumar J, Rajendran B, Bindhumadhava BS, Babu NSC (2017) XML wrapping attack mitigation using positional token. Proc Int Conf Public Key Infrastruct Appl (PKIA):36–42

33. Weiser S, Werner M, Brasser F, Malenko M, Mangard S, Sadeghi AR (2019) TIMBER-V: tag-isolated memory bringing fine-grained enclaves to RISC-V. Proc NDSS:1–15

34. Backes M, Bugiel S, Derr E (2016) Reliable third-party detection in android and its security applications. Proc ACM SIGSAC Conf Comput Commun Secur (CCS):356–367

35. Miessler D. 100 mb password dictionary. [Online]. Available: https://github.com/danielmiessler/SecLists/tree/master/Passwords

36. Lu RX, Cao ZF (2007) Simple three-party key exchange protocol. Comput Secur 26(1):94–97

37. Nam J, Paik J, Kang HK, Kim UM, Won D (2009) An off-line dictionary attack on a simple three-party key exchange protocol. IEEE Commun Lett 13(3):205–207

38. Roy A, Memon N, Togelius J, Ross A (2018) Evolutionary methods for generating synthetic MasterPrint templates: dictionary attack in fingerprint recognition. Proc Int Conf Biometrics (ICB):39–46

39. Cassola A, Robertson WK, Kirda E, Noubir G (2013) A practical, targeted, and stealthy attack against WPA enterprise authentication. Proc NDSS:1–15

40. Bhargavan K, Leurent G (2016) Transcript collision attacks: Breaking authentication in TLS, IKE, and SSH. Proc Netw Distrib Syst Secur Symp (NDSS):1–17

41. Vanhoef M, Piessens F (2017) Key reinstallation attacks: forcing nonce reuse in WPA2. Proc ACM SIGSAC Conf Comput Commun Secur (CCS):1313–1328

42. Vanhoef M, Piessens F (2018) Release the Kraken: New KRACKS in the 802.11 standard. Proc ACM SIGSAC Conf Commun Secur (CCS):299–314

43. Tu GH, Li Y, Peng C, Li CY, Wang H, Lu S (2014) Control-plane protocol interactions in cellular networks. Proc ACM Conf SIGCOMM 44:223–234

44. Rupprecht D, Jansen K, Popper C (2016) Putting LTE security functions to the test: a framework to evaluate implementation correctness. USENIX Association, Austin

45. Hussain SR, Chowdhury O, Mehnaz S, Bertino E (2018) LTEInspector: a systematic approach for adversarial testing of 4G LTE. Proc NDSS:1–15

46. Chen EY, Pei Y, Chen S, Tian Y, Kotcher R, Tague P (2014) OAuth demystified for mobile applicatoin developers. Proc ACM SIGSAC Conf Comput Commun Secur:892–903

47. Fernandes E, Jung J, Prakash A (2016) Security analysis of emerging smart home applications. Proc IEEE Symp Secur Privacy (SP):636–654

48. Jia Y, Xiao Y, Yu J, Cheng X, Liang Z, Wan Z (2018) A novel graph-based mechanism for identifying traffic vulnerabilities in smart home IoT. Proc IEEE Conf Comput Commun (INFOCOM):1493–1501

49. Pinkas B, Sander T (2002) Securing passwords against dictionary attacks. Proc 9th ACM Conf Comput Commun Secur (CCS):161–170

50. Mulliner C, Borgaonkar R, Stewin P, Seifert JP (2013) SMS-based one-time passwords: attacks and defense. Proc Int Conf Detection Intrusions Malware Vulnerabililty Assesment:150–159

51. Wang D, Ming J, Chen T, Zhang X, Wang C (2018) Cracking IoT device user account via brute-force attack to SMS authentication code. Proc 1st Workshop Radical Exp Secur:57–60

52. Liu Y (2015) Defense of WPA/WPA2-PSK brute forcer. Proc 2nd Int Conf Inf Sci Control Eng:185–188

53. Noh J, Kim J, Kwon G, Cho S (2016) Secure key exchange scheme for WPA/WPA2-PSK using public key cryptography. Proc IEEE Int Conf Consum Electron Asia (ICCE-Asia):1–4

54. Bhargavan K, Blanchet B, Kobeissi N (2017) Verified models and reference implementations for the TLS 1.3 standard candidate. Proc IEEE Symp Secur Privacy (SP):483–502

55. Duan X, Wang X (2016) Fast authenticatoin in 5G HetNet through SDN enabled weighted secure-context-information transfer. Proc IEEE Int Conf Commun (ICC):1–6

56. Zhao C, Huang L, Zhao Y, Du X (2017) Secure machine-type communications toward LTE heterogeneous networks. IEEE Wirel Commun 24(1):82–87

57. Yang R, Lau WC, Shi S (2017) Breaking and fixing mobile app authentication with OAuth2.0-based protocols. Proc Int Conf Appl Cryptogr Netw Secur:313–335

58. Shehab M, Mohsen F (2014) Securing OAuth implementations in smart phones. Proc 4th ACM Conf Data Appl Secur Privacy (CODASPY):167–170

59. Tian Y et al (2017) SmartAuth: user-centered authorization for Internet of Things. Proc 26th USENIX Secur Symp (USENIX Security):361–378

60. Celik ZB et al (2018) Sensitive information tracking in commodity IoT. Proc 27th USENIX Secur Symp (USENIX Security):1687–1704

61. Bastys I, Balliu M, Sabelfeld A (2018) If this then what?: controlling flows in IoT apps. Proc ACM SIGSAC Conf Comput Commun Secur (CCS):1102–1119

62. Celik ZB, Tan G, McDaniel P (2019) IoTGuard: dynamic enforcement of security and safety policy in commodity IoT. Proc NDSS:1–15

63. Jia YJ et al (2017) ContexIoT: towards providing contextual integrity to appified IoT platforms. Proc NDSS

64. Schuster R, Shmatikov V, Tromer E (2018) Situational access control in the Internet of Things. Proc ACM SIGSAC Conf Comput Commun Secur (CCS):1056–1073

65. Wang J, Chen Q (2014) ASPG: generating android semantic permissions. Proc IEEE 17th Int Conf Comput Sci Eng:591–598

66. Fernandes E, Paupore J, Rahmati A, Simionato D, Conti M, Prakash A (2016) FlowFence: practical data protection for emerging IoT application frameworks. Proc 25th USENIX Secur Symp (USENIX Security):531–548

67. Sikder AK, Aksu H, Uluagac AS (2017) 6thSense: a context-aware sensor-based attack detector for smart devices. Proc 26th USENIX Secur Symp (USENIX Security):397–414

68. Aafer Y, Tao G, Huang J, Zhang X, Li N (2018) Precise android API protection mapping derivation and reasoning. Proc ACM SIGSAC Conf Comput Commun Secur:1151–1164

An Intelligent Facial Expression Recognizer Using Modified ResNet-110 Using Edge Computing

Wenle Xu and Dimas Lima

1 Introduction

Human beings understand the world through five organs: sight, taste, hearing, smell, and touch. In terms of human sight, we call the information seen by the eyes image information. In the past, facial expression recognition can be done with a single sense or with the cooperation of multiple senses. It results from the joint action of global recognition and feature recognition. Facial expressions play the leading role in understanding the emotional state of a particular object. A large part of human emotions is observed on their faces. Therefore, in order to promote the exploitation of man-machine interaction, we build a facial expression recognition system in a situation [1]. The cornerstone of this recognizer is ResNet-110.

Knowing from the study that the purpose of Intelligent computing (IC) [2–4] is to attract intelligence, reasoning, sense, intelligence gathering, and decompose to computer systems. In the image processing, it is tough to extract useful information from source images manually. But neural networks can solve this problem. The biggest advantage of convolution neural networks (CNNs) [5–7] and many network models are that the row data is directly used as input. This is why CNNs is widely known in the area of image deep learning [8]. CNNs be made up of several convolution modules and few fully connected modules. The main purpose of convolutional module setting is to extract useful information from the training

W. Xu
School of Computer Science and Technology, Henan Polytechnic University, Jiaozuo, Henan, P.R. China
e-mail: xwl@home.hpu.edu.cn

D. Lima (✉)
Department of Electrical Engineering, Federal University of Santa Catarina, Florianópolis, Brazil
e-mail: dimaslima@ieee.org

© The Author(s), under exclusive license to Springer Nature Switzerland AG 2023
G. Srivastava et al. (eds.), *Security and Risk Analysis for Intelligent Edge Computing*, Advances in Information Security 103,
https://doi.org/10.1007/978-3-031-28150-1_11

set. In the highest convolution module, the convolution filter finally reflects the representative characteristics of the target. The outputs of several filters in the convolution module are classified according to the fully connected module [1].

The ordinary process of facial expression recognition is to take an expression image as the input of the recognizer, process the image through the recognizer, and finally obtain the classification result. According to the classification of facial expressions in reference [9], we also divided facial expressions into seven labels.

The detailed manufacture of facial expression recognition is zoned as four stages: In the first stage, our task is to take images. In the second stage, we simply preprocess these images. In the third phase, the designed model is used for feature extraction. Finally, the results are obtained. Many scholars have done relevant study. For example, Ali et al. [10] adopted the radon transform (RT) and traditional SVM algorithm. Lu et al. [11] employed the Haar wavelet transform (HWT) approach. With the development of convolutional neural networks and computer hardware, Ivanovsky et al. [12] used a convolutional neural network on GPU for feature extraction. Hasani et al. [13] combine modified Inception-ResNet layers with Conditional Random Fields to improve facial emotion recognition accuracy greatly. Yang [14] used cat swarm optimization (CSO) method and achieved a 89.49% accuracy. Li [15] employed biogeography-based optimization (BBO).

Later, Nwosu et al. [16] devised a dual-channel convolutional neural network. The first channel inputs extracted eye features. The second channel inputs extracted mouth features. After the above two features are fused and input into the complete connection layer, the human accuracy is improved. Li et al. developed ResNet-18 [17] and ResNet-50 [18] for expression classification using ResNet as a skeleton network. The above methods facilitate the effect of model.

However, we can find that most methods for facial emotion are unstable. Because we are easy to lose the original emotional information in translation operation. What's more, network models all have poor generalization and weak robustness. Faced with these issues, the main ways of feature extraction include: based on geometric, overall statistical, frequency domain, motion feature extraction. The most commonly method is based on frequency domain feature extraction. The main idea is to transform the image from the spatial domain to the frequency domain to extract its features (lower-level features). The essential method is Gabor wavelet transform. Motion feature extraction is the focus of future research.

At the heart of this study, we propose a exquisite facial expression recognition algorithm based on ResNet. The innovations of the chapter are as follows:

(i) An architecture called Revised ResNet-110 is devised.
(ii) Again, the weights are trained on the expression data.
(iii) The metrics of this system outperform the best approaches.

With the development of the Internet of things, the concept of edge computing has been rapidly popularized. In general, edge computing is part of the original server computing, but now it is calculated directly by the device. The server only needs the result and does not need the data of the process. In this chapter, our core idea

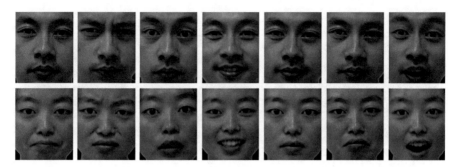

Fig. 1 Samples of our dataset

is to use some of the network's parameters to train a new network. Therefore, our method can be applied to edge computation.

2   Dataset

We drew on dataset [19]. The dataset contains seven categories of expression: happy, sadness, fear, anger, surprise, disgust, and neutral. Each category has 100 samples, and the total number of samples is 700. Figure 1 shows a male sample and a female sample.

3   Methodology

3.1   Convolution Modules

The detailed operation steps are as follows:

1. Use a filter [20] at a specific location in the image.
2. The weight value in the filter is multiplied by the pixel corresponding to the position in the input image.
3. Sum the above results, and the final value is the target pixels [21, 22].
4. Repeat steps 1–3 step until you have traversed all pixels.

The process is shown in Fig. 2.

According to the specific process above, we take the receptive field in the upper left corner as an example, and the strict operation steps of convolution can be obtained as follows:

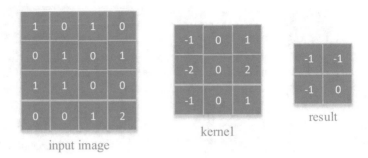

Fig. 2 Convolution process

$$1 \times (-1) + 0 \times 0 + 1 \times 1 + 0 \times (-2) + 1 \times 0 + 0 \times 2 + 1$$
$$\times (-1) + 1 \times 0 + 0 \times 1 = -1 \tag{1}$$

The summation idea of convolution says that pixels are summed by weights, and these weights can be learned in the process of back propagation. In the process of forward propagation, we need to initialize them.

The 3×3 filter we used in the previous article is usually called the Sobel filter. It is also called an edge detector. The specific use of convolution can be previewed in Refs. [23–25]: the brighter pixels in the output image represent the edge information in the original image. Recall the MNIST handwritten number classification problem. CNN trained on MNIST can find a specific number. Finding the number 1, for example, can be done by using edge detection to find two vertical edges that stand out from the image.

3.2 Max Pooling

The output pixels of adjacent pixels with very similar values often show similar values [26–28], and these values contain a lot of complicated and redundant information [29–31].

The pooling operation solves this problem. Pooling layer reduces the dimension of output values. Pooling is the search for a maximum or mean value in the surrounding pixels. Here is an example of Max pooling with pool size 2. The process is shown in Fig. 3.

Fig. 3 Max pooling process

Fig. 4 ReLU function

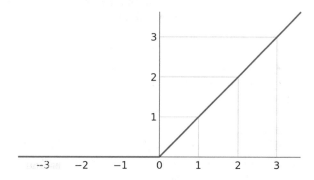

3.3 *Activation Function*

The immediate reason for adding a nonlinear activation function is to give the neuron a nonlinear relationship [32]. If this step is not introduced, the final result is a variant of the original perceptron [33].

Therefore, activation function are extremely important in the idea of deepening the neural network. We adopted the ReLU function in network model. The function change curve and function shown in Fig. 4 and Eq. (2)

$$l = \max(0, m) \tag{2}$$

In Eq. (2), m indicates the crop of the upper layer, l indicates the value through RELU. In Fig. 4, the level represents m and the standing represents l.

Among other things, RELU prevents gradient disappearance during backpropagation. In addition, according to the derivative of RELU, the convergence speed is very fast.

3.4 Modified ResNet-110

In theory, when we face complex problems, the deeper the network is, the better the performance is. However, it is found that with the deepening of the network, the accuracy of the training set decreases. The reason for this phenomenon is that the vanishing gradient. When the network reaches a certain depth, network training performance will degrade so that the accuracy will decrease. The residual module can solve this problem in ResNet. Because the residual module builds a direct connection and shortcut connection between input and output, the output needs to extract some new feature based on the original input. A building block is shown Fig. 5.

Figure 5 shows that ResNet module has two routes, identity route, and residual route. In the optimal state, the input x is optimal, then the residual mapping is going to be 0. The deepening of the network will greatly promote the future extraction of the network.

The backbone network in this chapter is Modified ResNet-110. We regard it as the final recognizer.

Officially, denoting the original mapping as H(x), denoting the residual mapping as F(x). We use nonlinear layers to configure the formula F(x) := H(x) − x. So after this process H(x) = F(x) + x. If our special case above is satisfied, at this time H(x) = x.

The first layer in the Modified ResNet-110 structure is convolution module. Then we use a stack of 6n layers with convolution operation on the feature maps of sizes (32, 16, 8), respectively, with 2n layers for each feature map size [34–37]. The size of all convolution kernels is 3 * 3. The corresponding number of filters is (16, 32, 64). Finally, perform an average pooling module, a 7-way fully-connected module, and softmax function. There are w stacked weighted layers [38].

Fig. 5 The building block

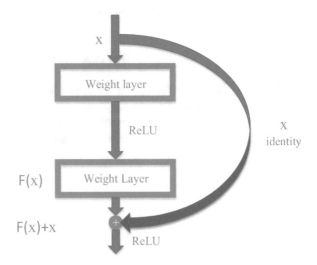

$$w = 6n + 2 \tag{3}$$

We explore $n^\star = 18$ that leads to a 110-layer Modified ResNet. Here $n^\star$ means the optimal n chosen in this work.

Suppose training error rate is ε, the learning rate is r, we have:

$$r = \begin{cases} 0.1 & \varepsilon < 80\% \\ 0.01 & \text{otherwise} \end{cases} \tag{4}$$

3.5 The Method of Training Weights

The loss function is as follows

$$L(w, b) = \sum_{i=1}^{n} (y_i - f(w, b))^2 \tag{5}$$

where y_i indicates the true label, and f indicates the output of model. In order to obtain the best model, our goal should be to maximize the total loss. According to the principle of gradient descent, in the process of back propagation, we use the following two equations for calculation.

$$w_{i+1} = w_i - \alpha * \frac{dL}{dw_i} \tag{6}$$

$$b_{i+1} = b_i - \alpha * \frac{dL}{db_i} \tag{7}$$

The function of the first few layers of the convolution network is to extract image features and not classify them until the last fully connected layer. We will directly use the first 109 layers of the ResNet-110 trained on CIFAR-10, and then use facial data to train the new full connection layer to complete the classification task. In this way, we can save training time.

4 Result and Discussions

4.1 Result Analysis

Table 1 reveals the sensitivity data for each category. Figure 6 reveals the curve of sensitivity change. As shown in Table 1 and Fig. 6, we can acquire that the sensitivity

Table 1 Result analysis on the sensitivity of each class

	Anger	Disgust	Fear	Happy	Neutral	Sadness	Surprise
Run1	97.14	97.14	94.29	97.14	98.57	95.71	97.14
Run2	95.71	94.29	97.14	94.29	94.29	95.71	94.29
Run3	98.57	97.14	100.00	100.00	97.14	97.14	94.29
Run4	98.57	94.29	95.71	94.29	98.57	97.14	94.29
Run5	98.57	97.14	92.86	94.29	98.57	100.00	97.14
Run6	98.57	100.00	98.57	98.57	100.00	98.57	95.71
Run7	92.86	95.71	90.00	97.14	97.14	92.86	97.14
Run8	92.86	95.71	92.86	92.86	98.57	98.57	92.86
Run 9	94.29	94.29	92.86	92.86	95.71	100.00	95.71
Run10	90.00	97.14	98.57	100.00	98.57	97.14	95.71
Average	95.72 ± 3.10	96.29 ± 1.81	95.29 ± 3.23	96.14 ± 2.78	97.71 ± 1.68	97.28 ± 2.18	95.43 ± 1.47

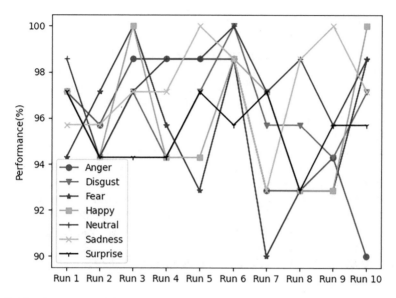

Fig. 6 The change curve of the sensitivities of each class

Table 2 Result analysis on the overall accuracies

Run	OA
1	96.86
2	94.57
3	97.71
4	96.57
5	96.71
6	98.71
7	95.00
8	94.14
9	95.86
10	96.71
Average	96.29 ± 1.42

of each emotion is: $95.72 \pm 3.10\%$ (anger), $96.29 \pm 1.81\%$ (disgust), $95.29 \pm 3.23\%$ (fear), $96.14 \pm 2.78\%$ (happy), $97.71 \pm 1.68\%$ (neutral), $97.28 \pm 2.18\%$ (sadness), $95.43 \pm 1.47\%$ (surprise). From this data we can get: that the sensitivity of neutral emotion is most big. Namely, the neutral emotion can be directly recognized. The second most recognizable expression is sadness. Disgust was a close second. From Table 2, we get that the OA (overall average accuracy) of the model after ten deals is $96.29 \pm 1.42\%$.

4.2 Comparison with Advanced Methods

The main current methods are HWT [11], CSO [14], and BBO [15], Table 3 shows
the indicator OA. The OA of match is shown in Table 3. It can be clearly see from
the data in Fig. 7 that the "Modified ResNet-110" model achieves an accuracy rate
of 96.29 ± 1.42%. The accuracy of BBO [15] was 93.79%, ranking second. The
accuracy of CSO was 89.49%, ranking third. The accuracy of HWT was 78.37%,
ranking the last.

From Table 1 we can infer that the highest OA obtained by "Modified ResNet-
110" method mostly lies on (i) the ability of CNN to analyze images; (ii) the network
deepens to extract advanced features.

The second is the BBO method, which is initially a swam intelligent optimization
algorithm. The heart of BBO algorithm is divided into two modules, migration and
mutation. The third method is CSO, which is a global optimization algorithm. The
essential idea of CSO is to simulate cat behavior.

We can deepen the fun of ResNet by continuing to explore it.

Table 3 Comparison with advanced methods

Method	OA
HWT [11]	78.37 ± 1.50
CSO [14]	89.49 ± 0.76
BBO [15]	93.79 ± 1.24
Modified ResNet-110	96.29 ± 1.42

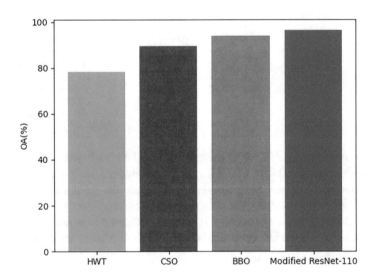

Fig. 7 OA of advanced methods

Table 4 Comparison with other ResNet variants

Method	OA
ResNet-18 [17]	94.80 ± 1.43
ResNet-50 [18]	95.39 ± 1.41
Modified ResNet-110	96.29 ± 1.42

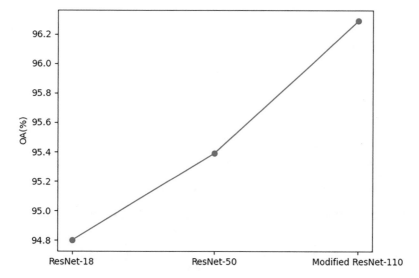

Fig. 8 The change curve of OA of ResNet transformation

4.3 *Comparison with Other ResNet Transformation*

Li, et al. utilized ResNet-18 [17] and ResNet-50 [18] of ResNet transformation as recognizer. The accuracy rates were achieved 94.80 ± 1.43% and 95.39 ± 1.41%, alone. According to Table 4 and Fig. 8, our proposed model fulfils upper accuracy.

One promising solution is to improve swarm intelligence by tweaking parameters. In the following tasks, we boldly test particle swarm optimization [39–42] and else global optimization methods. The essential step of Software-defined networking (SDN) [43–46] is the network mode based on software controller controllers or application programming interfaces. In the immediate future, we must associate SDNs with this task.

5 Conclusion

This chapter designs a recognizer based on ResNet-110 as skeleton network for facial expression. Compared with the current mainstream models, we verify that the accuracy of the model is greatly improved. The direct reason for the low classification accuracy may be the small number of samples in the training set.

Our further task will focus on directions. The first direction is that using additional data from other datasets or data enhancements to the corresponding categories may improve the performance of our CNN model.

In the second direction, we continually deepen the convolution network to promote accuracy. What's more, we can collect more data to train the model better and optimize the hyper-parameter of the network.

References

1. Jeon J, Park JC, Jo YJ, Nam CM, Kim DS (2016) A real-time facial expression recognizer using deep neural network. In: The 10th international conference
2. Karaca Y (2022) Secondary pulmonary tuberculosis recognition by rotation angle vector grid-based fractional Fourier entropy. Fractals 30:2240047
3. Satapathy SC (2022) Secondary pulmonary tuberculosis identification via pseudo-Zernike moment and deep stacked sparse autoencoder. J Grid Comput 20:1
4. Govindaraj V (2022) Deep rank-based average pooling network for COVID-19 recognition. Comput Mater Contin 70:2797–2813
5. Elamin A, El-Rabbany A (2022) UAV-based multi-sensor data fusion for urban land cover mapping using a deep convolutional neural network. Remote Sens 14:4298
6. Luo Y, Zeng YC, Lv RZ, Wang WH (2022) Dual-stream VO: visual odometry based on LSTM dual-stream convolutional neural network. Eng Lett 30
7. Laishram A, Thongam K (2022) Automatic classification of oral pathologies using orthopanto-mogram radiography images based on convolutional neural network. Int J Interact Multimedia Artif Intell 7:69–77
8. Zhang Y-D, Dong Z-C (2020) Advances in multimodal data fusion in neuroimaging: overview, challenges, and novel orientation. Inf Fusion 64:149–187
9. Ekman P, Friesen W (1978) Facial action coding system: a technique for the measurement of facial movement. Consulting Psychologists Press, Palo Alto
10. Ali H, Hariharan M, Yaacob S, Adom AH (2015) Facial emotion recognition based on higher-order spectra using support vector machines. J Med Imaging Health Inform 5:1272–1277
11. Lu S, Evans F (2017) Haar Wavelet transform based facial emotion recognition. In: 2017 7th international conference on education, management, computer and society (EMCS 2017)
12. Ivanovsky L, Khryashchev V, Lebedev A, Kosterin I (2017) Facial expression recognition algorithm based on deep convolution neural network. In: 2017 21st conference of Open Innovations Association (FRUCT)
13. Hasani B, Mahoor MH (2017) Spatio-temporal facial expression recognition using convolutional neural networks and conditional random fields. In: IEEE, pp 790–795
14. Yang W (2017) Facial emotion recognition via discrete wavelet transform, principal component analysis, and cat swarm optimization. Lect Notes Comput Sci 10559:203–214
15. Li X (2020) Facial emotion recognition via stationary wavelet entropy and biogeography-based optimization. EAI Endorsed Trans e-Learn 6:e4
16. Nwosu L, Hui W, Jiang L, Unwala I, Zhang T (2018) Deep convolutional neural network for facial expression recognition using facial parts. In: 2017 IEEE 15th intl conf on dependable, autonomic and secure computing, 15th intl conf on pervasive intelligence and computing, 3rd intl conf on big data intelligence and computing and cyber science and technology congress (DASC/PiCom/DataCom/CyberSciTech)

17. Li B (2021) Facial expression recognition via ResNet-18. In: Lecture notes of the institute for computer sciences, social informatics and telecommunications engineering, vol 388, pp 290–303
18. Li B (2021) Facial expression recognition via ResNet-50. Int J Cogn Comput Eng 2:57–64
19. Lu HM (2016) Facial emotion recognition based on biorthogonal wavelet entropy, fuzzy support vector machine, and stratified cross validation. IEEE Access 4:8375–8385
20. Bazai H, Kargar E, Mehrabi M (2021) Using an encoder-decoder convolutional neural network to predict the solid holdup patterns in a pseudo-2d fluidized bed. Chem Eng Sci 246:116886
21. Hsu CF, Chien TW, Yan YH (2021) An application for classifying perceptions on my health bank in Taiwan using convolutional neural networks and web-based computerized adaptive testing: a development and usability study. Medicine 100:e28457
22. Nguyen NHT, Perry S, Bone D, Le HT, Nguyen TT (2021) Two-stage convolutional neural network for road crack detection and segmentation. Expert Syst Appl 186:115718
23. Wang S-H (2021) DSSAE: deep stacked sparse autoencoder analytical model for COVID-19 diagnosis by fractional Fourier entropy. ACM Trans Manag Inf Syst 13:2–20
24. Zhu Z (2021) PSCNN: PatchShuffle convolutional neural network for COVID-19 explainable diagnosis. Front Public Health 9:768278
25. Khan MA (2021) Pseudo Zernike moment and deep stacked sparse autoencoder for COVID-19 diagnosis. CMC Comput Mater Contin 69:3145–3162
26. Parcham E, Ilbeygi M, Amini M (2021) CBCapsNet: a novel writer-independent offline signature verification model using a CNN-based architecture and capsule neural networks. Expert Syst Appl 185:115649
27. Sitaula C, Shahi TB, Aryal S, Marzbanrad F (2021) Fusion of multi-scale bag of deep visual words features of chest X-ray images to detect COVID-19 infection. Sci Rep 11:23914
28. Zwingmann L, Zedler M, Kurzner S, Wahl P, Goldmann JP (2021) How fit are special operations police officers? A comparison with elite athletes from Olympic disciplines. Front Sports Active Living 3:742655
29. Jiang X (2021) Multiple sclerosis recognition by biorthogonal wavelet features and fitness-scaled adaptive genetic algorithm. Front Neurosci 15:737785
30. Zhang Z, Zhang X (2021) MIDCAN: a multiple input deep convolutional attention network for Covid-19 diagnosis based on chest CT and chest X-ray. Pattern Recogn Lett 150:8–16
31. Wang S-H (2021) SOSPCNN: structurally optimized stochastic pooling convolutional neural network for tetralogy of fallot recognition. Wirel Commun Mob Comput 2021:5792975
32. Fushimura Y, Hoshino A, Furukawa S, Nakagawa T, Hino T, Taminishi S et al (2021) Orotic acid protects pancreatic 13 cell by p53 inactivation in diabetic mouse model. Biochem Biophys Res Commun 585:191–195
33. Karadurnius E, Goz E, Taskin N, Yuceer M (2020) bromate removal prediction in drinking water by using the least squares support vector machine (LS-SVM). Sigma J Eng Nat Sci (Sigma Muhendislik Ve Fen Bilimleri Dergisi) 38:2145–2153
34. Ahn Y, Hwang JJ, Jung YH, Jeong T, Shin J (2021) Automated mesiodens classification system using deep learning on panoramic radiographs of children. Diagnostics 11:1477
35. Ananda A, Ngan KH, Karabag C, Ter-Sarkisov A, Alonso E, Reyes-Aldasoro CC (2021) Classification and visualisation of normal and abnormal radiographs; a comparison between eleven convolutional neural network architectures. Sensors 21:5381
36. Narin A, Isler Y (2021) Detection of new coronavirus disease from chest x-ray images using pre-trained convolutional neural networks. J Fac Eng Archit Gazi Univ 36:2095–2107
37. Bukhari SUK, Mehtab U, Hussain SS, Syed A, Armaghan SU, Shah SSH (2021) The assessment of deep learning computer vision algorithms for the diagnosis of prostatic adenocarcinoma diagnosis of prostatic adenocarcinoma using computer vision. Ann Clin Anal Med 12:31–34
38. He K, Zhang X, Ren S, Sun J (2016) Deep residual learning for image recognition. In: 2016 IEEE conference on computer vision and pattern recognition (CVPR), pp 770–778
39. Hou X-X (2017) Alcoholism detection by medical robots based on Hu moment invariants and predator-prey adaptive-inertia chaotic particle swarm optimization. Comput Electr Eng 63:126–138

40. An JP, Li XH, Zhang ZB, Man WX, Zhang GH (2020) Joint trajectory planning of space modular reconfigurable satellites based on kinematic model. Int J Aerosp Eng 2020:8872788–8872717

41. Delavari H, Naderian S (2020) Design and HIL implementation of a new robust fractional sliding mode control of microgrids. IET Gener Transm Distrib 14:6690–6702

42. Etedali S (2020) Ranking of design scenarios of TMD for seismically excited structures using TOPSIS. Front Struct Civ Eng 14:1372–1386

43. Ghosh U, Chatterjee P, Shetty S (2022) Securing SDN-enabled smart power grids: SDN-enabled smart grid security. In: Research anthology on smart grid and microgrid development. IGI Global, pp 1028–1046

44. Ghosh U, Chatterjee P, Shetty S, Datta R (2020) An SDN-IoT– based framework for future smart cities: addressing perspective. In: Internet of Things and secure smart environments. Chapman and Hall/CRC, pp 441–468

45. Ghosh U, Chatterjee P, Tosh D, Shetty S, Xiong K, Kamhoua C (2017) An SDN based framework for guaranteeing security and performance in information-centric cloud networks. In: 2017 IEEE 10th international conference on cloud computing (CLOUD), pp 749–752

46. Ghosh U, Datta R (2015) A secure addressing scheme for large-scale managed MANETs. IEEE Trans Netw Serv Manag 12:483–495

Blockchain Based Simulated Virtual Machine Placement Hybrid Approach for Decentralized Cloud and Edge Computing Environments

Suresh Rathod, Rahul Joshi, Sudhanshu Gonge, Sharnil Pandya, Thippa Reddy Gadekallu, and Abdul Rehman Javed

1 Introduction

In recent years, cloud computing gaining popularity because of virtualization. Virtualized resources are deployed, provisioned, and released with minimal management effort [1–19]. Virtualization addresses varying resource requirements by incorporating partitioning, isolation, and encapsulation [20–27, 41, 42]. Virtual Machine (VM) is a core element in a cloud environment that runs on top of the hypervisor and do utilize in the derlying server's resources. Each VM differs from other VM by its assigned resources, the CPU architecture [28–30], configured operating system, attached storage type, network utilization, and the job [31–40,

The original version of the chapter has been revised. A correction to this chapter can be found at
https://doi.org/10.1007/978-3-031-28150-1_14

S. Rathod · R. Joshi · S. Gonge · S. Pandya
Symbiosis Institute of Technology, Pune, Symbiosis International University (Deemed
University), Pune, India
e-mail: suresh.rathod@sitpune.edu.in; rahulj@sitpune.edu.in; sudhanshu.gonge@sitpune.edu.in;
sharnil.pandya@sitpune.edu.in

T. R. Gadekallu (✉)
School of Information Technology and Engineering, Vellore Institute of Technology, Vellore,
Tamil Nadu, India

Department of Electrical and Computer Engineering, Lebanese American University, Byblos,
Beirut, Lebanon
e-mail: thippareddy.g@vit.ac.in

A. R. Javed
Air University, Islamabad, Pakistan
e-mail: abdulrehman.cs@au.edu.pk

223
G. Srivastava et al. (eds.), *Security and Risk Analysis for Intelligent Edge
Computing*, Advances in Information Security 103,
https://doi.org/10.1007/978-3-031-28150-1_12

43–47] it executes. As a result, servers in DC have multiple VMs running on it with different job completion times. Each server in the data centre fulfils varying resource demands if it does not reach its resource threshold usage limit. The static threshold limit on underlying servers' resources degrades the server's performance, which can be improved by migrating the task from the VM or migrating the VM to one of the servers located in the data centre. In task migration, the tasks from the current VM migrated to other VM, running on the same or different servers in the data centre. The VM migration involves the migration state of the VM and the associated memory pages of the VM to another server. Several cloud providers like Google, Amazon, HP, and IBM provide services by adopting either centralized or de-centralized cloud architecture [77–80]. In a centralized cloud, the central entity server is configured with the agent that manages and controls the resources of servers running in the data centre. In contrast, the rest servers in the data centre are configured to share resource details with the agent running on the central server. Here, in the centralized cloud architecture, the central entity can decide to manage resources running on remote servers located in the same data centres. In the decentralized cloud and edge computing environment [81, 82], each host is configured with the agents to collect and share the underlying server's resource details with the peer servers. Each peer server in the decentralized cloud and edge computing environment is configured with its upper threshold for resource usage. Suppose the peer server finds its resource usage is reached with its upper threshold. In that case, it does initiate a task migration process or VM migrating to one of its peer servers considering the peer servers' utilization Table. VM contains the user's sensitive information or applications data, which VM might have used to access service hosted on the same cloud or get accessed from the server deployed on cloud or the third party server. In migration, there are chances to steal the VM's data while moving to the destination server. This requires the necessary attention and action to avoid loss of information while data is in transit, at the same time availability of the server where VM gets placed by the peer server. To do so the is paper discusses the blockchain-based adaptive VM placement hybrid approach for decentralized cloud and edge computing environment and edge computing environment, where it discusses how blockchain could be used in decentralized cloud and edge computing environment to ensure availability of the peer server after VM placement on it.

2   Related Work

Grigorenko et al. (2016) [4] have discussed energy-based VM placement. In this, authors have considered penalty cost and energy consumption as the parameters during VM selection. The solution proposed by the authors [4] suffers from performance degradation; that is, if there is an increase in energy cost and the penalty cost, the overall performance would be degraded [4]. Benali et al. (2016) [24] have proposed CPU utilization-based distributed load balancing and have proposed hypercube-based VM placement. In this work, each server takes decisions for VM migration without considering the destination server's future

CPU utilization. Bagheri Z and Zamanifar K [25] have proposed an optimum dynamic VM Placement policy [25] considering the peer server's CPU utilization. In this work, authors have achieved VM migration with maximum processing power (MPP) [25], and VM placement with random server selection (RS). The authors have discussed how VM can continue to use its firewall rules after it migrated to another server. Ferdaus et al. [26] discuss hierarchical Decentralized Dynamic VM Consolidation Framework [48], wherein they discussed the importance of global controller in decision-making, and how this global controller considers the server's future CPU utilization during the VM migration selection process. Pantazoglou et al. [49], the author has proposed VM placement using Ant Colony Optimization (ACO) technique. Here the author has hypercube- based VM migration for distributed load balancing using CPU utilization. A randomized probabilistic technique for distributed live VM migration proposed by Nikzad [50], discussed servers pair formation and initiating VM migration in the selected server pair. Zhao Y and Huang W [51] have proposed VM placement with correlation. In this paper, authors have considered centralized cloud architecture during their research. Fu X proposed Cluster-based VM consolidation and Zhou C [52], where they have discussed batch-oriented VM consolidation and on-demand VM placement. Teng F [53] discussed multi-target based VM placement using a genetics algorithm. Here authors have considered SLA violation and CPU utilization as the basis for VM consolidation in centralized cloud architecture. Arianyan E et al. [54], have proposed Reinforcement Learning-based VM placement. Considering the server traces, the authors discussed how the centralized server learns VM deployment and puts the server in sleep mode or inactive mode. Jayamala R and Valarmathi A [55], have proposed ED-VMM (Enhanced decentralized virtual machine migration) [56], considering a linear prediction model that reduces the server's energy utilization in the data centres. Here, the authors showed that after applying the EDVMM approach, the resource utilization is enhanced and reduced the number of virtual machine migration. Sagar D et al. [57], have proposed homogeneous load balancing using Table marriage matching problems and have achieved server load by exchanging VM's in between servers.

3   Decentralized Blockchain-Based Edge Computing Model

This section discusses proposed predictive VM migration in decentralized cloud and edge computing environment architecture. Figure 1 shows the generalized predictive model for the decentralized cloud and edge computing environment architecture. Each server in the proposed architecture is configured with the components, as shown in Fig. 1.

HC Resource Monitor (HCRM) Each server in the DC is configured with this component. This component stores peer server details in peer Tables and interacts with Virtual Host Manager (VHM) [58–60].

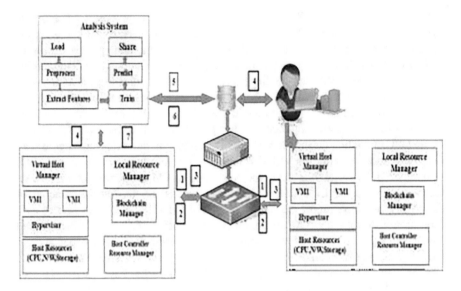

Fig. 1 Component diagram for the proposed framework

Local Resource Monitor (LRM) Each server has this component. This component is provisioned to interact with hyper-visor and to collect underlying server details. LRM forwards local servers' data with the HCRM [61–63].

Virtual Host Manager (VHM) This component interacts with HCRM and retrieves peer server details. This component performs the task of identifying a peer server for VM placement and initiating VM migration. VHM also calculates the local server's upper threshold and future CPU utilization and the server identified for VM placement. It calls the blockchain manager to add the specified server's details and the VM details to be migrated [64–66, 83–93].

Blockchain manager (BM) This is a component added as an agent to each server. This component interacts with HCRM and VHM. It adds VM's state details and the underlying server's details with the identified server.

Edge Computing (EC) Edge-computing enabled decentralized cloud processes data in an autonomous manner and transmits the processed data to the cloud such as Google Firebase [67].

4 Working of Proposed Model

Initially, each server is configured with the component mentioned above. Each server in the proposed architecture shares its details with the peers using the form below after a fixed interval of time. Here, the status flag specifies whether the server is underutilized or over-utilized at the current time instant. The address identifies the current IP address of the server, no. of VM, represents the number of running VM instances, the CPU utilization is the current CPU utilization [26, 68, 69].

The VHM at each server, observers the peer server's details, to find a suitable peer server, to whom it can move its current VM instances to reduce its utilization caused due to varying demands. Doing so initiates the process of finding a suitable server for VM placement and tires to reduce its CPU utilization upon calling DPVP [70–72]. DPVP finds the appropriate server and the details for the same. Details include current utilization of the identified server, future utilization of the identified server after VM is migrated, and the future utilization of the underlying server after VM gets migrated to another server. To do so, VHM calls DPVP, which computes the upper threshold of each peer server, finds the suitable server from peer servers, and initiates VM migrations on an appropriate identified server [73–76]. The DPVP algorithm is as shown in Algorithm 1.

The LRM at each server computes the current utilization of the underlying server using Eq. (1) and shares it with HCRM to update in the peer server's utilization Table.

$$H_U = \sum_{i=0}^{n} VM_i \tag{1}$$

Here HU is the server's utilization identified by u [20]. It is the sum of all virtual servers [20] VM_i running on the server u at time interval t.

Servers in data centres have heterogeneous configurations, each server varying resource threshold limit. The upper threshold, for each server computed with MAD [9], the equation for computing upper threshold is as in Eq. (3),

$$MAD = \sum_{t}^{n} \frac{y_t - \hat{y}}{n} \tag{2}$$

Here, y_t represent actual CH's utilization, and n represent the number of observations and $\hat{y}$ represent fitted value at time t.

$$Upper\ Threshold = 1 - MAD \tag{3}$$

DPVP at VHM initiates a process to find the server with minimum CPU utilization using the find in server. Here, peer server utilization is given as input to it. The algorithms find a server with minimum CPU utilization and return the address. The below algorithm is for finding a peer server that has minimum utilization.

Upon receiving a server with minimum CPU utilization, DPVP marks it as the destination server and initiates a process to find the future CPU utilization of the identified peer server. To do so, DPVP calls the DES algorithm and finds the future CPU utilization of the identified server, and upper threshold considering its CPU utilization available at the current instance with the HCRM's peer utilization Table.

$$S_t = \alpha y_t + (1 - \alpha)(s_t - 1 + b_t - 1), \quad 0 \le \alpha \le 1 \tag{4}$$

$$b_t = \gamma(s_t - s_t - 1) + (1 - \gamma)b_t - 1), \quad 0 \le \gamma \le 1 \tag{5}$$

Algorithm 1 DPVP Approach for Hybrid Decentralized Cloud and Edge Computing Environments

```
procedure DPVP(hostList,currUtil)
    maxUtilServe ←null
    hosts ←hostList
    med ←med[length(currUtil) ]
    median ← median[length(currUtil)]
    for each host in currUtil do
        if hostUtil.getvalue() > 0.7 then
            maxUtilServer.put(hostUtil.getKey(),hostUtil.getValue())
        end if
    end for
    for each host in maxUtilServer do
        srcAddress=host
        util=findUtilization(currUtil[host])
        while util <= 0.7 do
            util=findUtilization(hostList,currUtil)
            minServer=findMinServer(currUtil)
            vmut=findMinVMUtil(srcAddress,currUtil)
            setVmUtil(vmut)
            putil=doForecast(minServer)
            fhost=dofindFutureData()
            for each host in currUtil do
                if minServer == currUtil[host] then
                    for each serv in med do
                        if med[serv] == minServer then
                            address=med[serv].getKey()
                            if med[serv] == minServer then
                                medValue=med[serv].getValue()
                                if medValue > 0.9 then
                                    medValue=0.9
                                end if
                                if medValue < putil then
                                    address=findMinServer(currUtil)
                                    BM=Connect(address,srcAddress,currUtil)
                                    if BM == 1 then
                                        vm=findMinVM(srcAddress,currUtil)
                                        Formtunnel(srcAddress,address,util)
                                        migrate(srcAddress,address,vm)
                                        medValue=0.9
                                    else if BM ==0 then
                                        minServer=findNextMinServer(hostList)
                                        repeat step 34 to 39
                                    end if
                                else if medvalue >putil then
                                    vm=findMinVMUtil(srcAddress,util)
                                    Formtunnel(srcAddress,currUtil)
                                    migrate(srcAddress,address,vm)
                                end if  2
                            end if
                        end if
                    end for
                end if
            end for
        end while
    end for
end procedure
```

The equation to predict CPU utilization of the identified peer server is calculated using Eq. (6)

$$f_{t+m} = s_t + mb_t \tag{6}$$

Here S_t represents CH's smooth values [20] at time t, the y_t represents observed values over a time period t [20]. b_t represent trend factor over period t values for the previous [20] period b_{t-1}. This f_{t+m} called the smoothing function.

Upon finding future CPU utilization of the peer server, the VHM at the originator server initiates the process to find the adaptability of the identified peer server. Here, BM at the originator server, add's the underlying server's details in the blockchain block. Block contains the following information in it.

1. VM's current CPU utilization,
2. The status value represents whether to accept a block or reject block
3. Request info, specifying block contains information for requesting a VM to place or reply for VM placement.
4. The originator peer public key

Upon receiving this block, BM at the destination server checks flag value and requests info value. If it finds the VM cannot be placed on the destination server after seeing future CPU utilization of the underlying host before it is put on it. BM at identified destination peer server alters the value of the status flag from 0 to 1. Otherwise, it will keep as it is, and the request flag from 1 to 2 specifies a reply to the originator server after adding its public key to the block. This block gets validated by the miners; after this, it would be added to the chain. If BM finds multiple blocks for VM placement, it traverses each block and finds its adaptability one by one on a first-come by first server basis. It will discard rest blocks if it finds a suiTable VM that can be adopted.

The BM, at the originator server, upon receiving block from the peer server, checks the status flag. If it finds a value equal to 1, it starts searching a new server from the current peer utilization Table and repeats it until it finds a suiTable peer server for VM placement.

5 Result and Discussion

The proposed framework was developed considering distributed peer-to-peer network. Each server is configured with a blockchain client, which accesses blockchain and updates content to it. The VM's traces are applied to each server considering Azure VM traces [23].

Each server in the decentralized cloud and edge computing environment shares its utilization details with the peer servers after a fixed interval. Each server updates its peer servers details in its peer Table. The snippet for this is shown in Table 1. The VHM at each server calls HCRM, runs on the underlying server, and

Table 1 Current utilization
table

CH. Address	CPU Util	No. VM	Status
10.0.0.1	0.113	1	FALSE
10.0.0.2	0.125	2	FALSE
10.0.0.3	0.90	4	TRUE

starts monitoring the underlying server's current utilization and the peer server's utilization Table. If VHM finds the underlying server is getting over-utilized and needs utilization to be reduced, it calls to DPVP. DPVP at each server receives an updated list from the HCRM and starts finding the server with minimum CPU utilization. From Table 1, it was discovered that 10.0.0.3 has a current CPU utilization is 0.9. Upon finding out that the server has its utilization reached the maximum upper threshold, VHM calls DPVP to find a suiTable peer server where it can migrate its VM. To do so, the DPVP at the originator server initiates a process to find a server whose current CPU utilization is lower compared with all peer servers; after looking in peer servers' utilization Table 1 found that a server with address 10.0.0.1 has lower CPU utilization than the upper threshold of 0.9. Its current CPU utilization is lesser than the rest servers in the peer utilization Table. The VHM, after finding 10.0.0.01 has less CPU utilization, calls BM to initiate a process for VM migration. The BM at 10.0.0.3 logins with its credentials, and creates a block by adding its own server's public key, the address of the destination server (10.0.0.1), VM's current CPU utilization in the data field. The miner server in these architectures validates the 10.0.0.2 server's block and adds it to the blockchain chain [4]. BM wake-ups after the specified interval and checks the block in the chain specifying its credentials. If it finds the block is for it and it is a destination server, and the request is to adopt the VM, then it calls its local HCRM and finds its future utilization. If it finds its upper threshold limit and its future CPU utilization is below 0.9, it marks it can adopt this VM on it. After this, it updates the request field with the value equal to two and status as 0. If it finds it cannot take a load of new VM on it, the peer server can change the status value from 0 to 1, update request info to 2, and add its address. Figure 2 Simulated Blockchain traces during VM migrations.

Figure 2 shows the chained output. In Fig. 2, the first column represents VM's CPU utilization, the second block shows the status can be 0 or 1. Zero accepted, 1 for rejected. The third column represents the server that has requested or replied, 1 represents request, and 2 represents reply from peer server. The last column represents the server's public key.

6 Conclusion

The hybrid decentralized predictive blockchain-based VM migration avoids the overutilization of peer servers due to multiple VM placements on the server in decentralized cloud and edge computing environments. Restricting peer replies on a first come first basis helps the destination server to maintain its utilization below

Fig. 2 Blockchain output for the proposed hybrid approach for decentralized cloud and edge computing environments

the upper threshold in decentralized cloud and edge computing environments. VM placement communication among peer servers through blockchain helps tamper-proof communication. The use of DPVP with a two-threshold limit ensures each server has a dynamic threshold and reduces the number of active servers in the framework. In future, a generalized approach for Fog computing environment can be proposed considering privacy and security aspects.

References

1. National Institute of Standards and Technology NIST [online]. Available https://www.nist.gov/

2. Booth G, Soknacki A, Somayaji A (2013) Cloud security: attacks and current defenses, pp 56–62

3. Luo Y, Zhang B, Wang X, Wang Z, Sun Y, Chen H (2008) Live and incremental whole-system migration of virtual machines using block-bitmap. In: Proceedings – IEEE international conference on cluster computing, ICCC, vol Proceeding, pp 99–106

4. Ghayvat H, Pandya S, Bhattacharya P, Zuhair M et al (2021) CP-BDHCA: blockchain-based confidentiality-privacy preserving Big Data scheme for healthcare clouds and applications. IEEE J Biomed Health Inform (J-BHI). https://doi.org/10.1109/JBHI.2021.3097237

5. Pandya S, Sur A, Solke N (2021) COVIDSAVIOUR: a novel sensor-fusion and deep learning-based framework for virus outbreaks. Front Public Health. https://doi.org/10.3389/fpubh.2021.797808

6. Pandya S, Ghayvat H (2021) Ambient acoustic event assistive framework for identification, detection, and recognition of unknown acoustic events of a residence. Adv Eng Inform 47:1012

7. Grygorenko D, Farokhi S, Brandic I (2016) Cost-aware VM placement across distributed DCs using Bayesian networks. In: Lecture notes in computer science (including subseries Lecture notes in artificial intelligence and Lecture notes in bioinformatics), vol 9512, pp 32–48

8. Dong J, Jin X, Wang H, Li Y, Zhang P, Cheng S (2013) Energy-saving virtual machine placement in cloud data centers. In: Proceedings – 13th IEEE/ACM international symposium on cluster, cloud, and grid computing, CCGrid 2013, pp 618–624

9. Diaconescu D, Pop F, Cristea V (2013) Energy-aware placement of VMs in a datacenter. In: Proceedings – 2013 IEEE 9th international conference on intelligent computer communication and processing, ICCP 2013, September, pp 313–318

10. Dias DS, Costa HMK (2012) Online traffic-aware virtual machine placement in data center networks, pp 1–8

11. Daradkeh T, Agarwal A (2017) Distributed shared memory based live VM migration. In: IEEE international conference on cloud computing, CLOUD, pp 826–830

12. Majhi SK, Dhal SK (2016) A security context migration framework for Virtual Machine migration. In: 2015 international conference on computing and network communications, CoCoNet 2015, pp 452–456

13. Mehta P, Pandya S (2020) A review on sentiment analysis methodologies, practices and applications. Int J Sci Technol Res 9(2):601–609

14. Ghayvat H, Awais M, Gope P, Pandya S, Majumdar S (2021) ReCognizing SUspect and PredictiNg ThE SpRead of Contagion Based on Mobile Phone LoCation DaTa (COUNTERACT): a system of identifying COVID-19 infectious and hazardous sites, detecting disease outbreaks based on the internet of things, edge computing, and artificial intelligence. Sustain Cities Soc:102798

15. Mehbodniya A, Lazar AJ, Webber J, Sharma DK, Jayagopalan S, Singh P, Rajan R, Pandya S, Sengan S (2021) Fetal health classification from cardiotocographic data using machine learning, expert systems. Wiley

16. Ghayvat H, Pandya S, Bhattacharya P, Mohammad Z, Mamoon R, Saqib H, Kapal D (2021) CP-BDHCA: blockchain-based confidentiality-privacy preserving Big Data scheme for healthcare clouds and applications. IEEE J Biomed Health Inform 25:1–22

17. Mishra N, Pandya S (2021) Internet of Things applications, security challenges, attacks, intrusion detection, and future visions: a systematic review. IEEE Access

18. Mehta P, Pandya S (2021) Harvesting social media sentiment analysis to enhance stock market prediction using deep learning. PeerJ Comput Sci. https://doi.org/10.7717/peerj-cs.476

19. Ghayvat H, Awais M, Pandya S, Ren H, Akbarzadeh S, Chandra Mukhopadhyay S, Chen C, Gope P, Chouhan A, Chen W (2019) Smart aging system: uncovering the hidden wellness parameter for well-being monitoring and anomaly detection. Sensors 19(4):766

20. Awais M, Ghayvat H, Krishnan Pandarathodiyil A, Nabillah Ghani WM, Ramanathan A, Pandya S, Walter N, Saad MN, Zain RB, Faye I (2020) Healthcare Professional in the Loop (HPIL): classification of standard and oral cancer-causing anomalous regions of oral cavity using textural analysis technique in autofluorescence imaging. Sensors 20:5780

21. Patel CI, Labana D, Pandya S, Modi K, Ghayvat H, Awais M (2020) Histogram of oriented gradient-based fusion of features for human action recognition in action video sequences. Sensors 20(24):7299

22. Barot V, Kapadia V, Pandya S (2020) QoS enabled IoT based low cost air quality monitoring system with power consumption optimization. Cybern Inf Technol 20(2):122–140, Bulgarian Academy of Science

23. Pandya S, Wakchaure MA, Shankar R, Annam JR (2022) Analysis of NOMA-OFDM 5G wireless system using deep neural network. J Def Model Simul 19(4):799–806

24. Sur A, Sah R, Pandya S (2020) Milk storage system for remote areas using solar thermal energy and adsorption cooling. Mater Today 28(3)

25. Ghayvat H, Pandya S, Patel A (2020) Deep learning model for acoustics signal based preventive healthcare monitoring and activity of daily living. In: 2nd international conference on data, engineering and applications (IDEA), Bhopal, India, pp 1–7. https://doi.org/10.1109/IDEA49133.2020.9170666

26. Pandya S, Shah J, Joshi N, Ghayvat H, Mukhopadhyay SC, Yap MH (2016) A novel hybrid based recommendation system based on clustering and association mining. In: 2016 10th international conference on sensing technology (ICST), November. IEEE, pp 1–6

27. Karn AL, Pandya S, Mehbodniya A et al (2021) An integrated approach for sustainable development of wastewater treatment and management system using IoT in smart cities. Soft Comput:1–7

28. Pandya S, Thakur A, Saxena S, Jassal N, Patel C, Modi K, Shah P, Joshi R, Gonge S, Kadam K, Kadam P (2021) A study of the recent trends of immunology: key challenges, domains, applications, datasets, and future directions. Sensors 21:7786. https://doi.org/10.3390/s21237786

29. Reference

30. Wan X, Zhang X, Chen L, Zhu J (2012) An improved vTPM migration protocol based trusted channel. In: 2012 international conference on systems and informatics, ICSAI 2012, No. ICSAI, pp 870–875

31. Ahmad N, Kanwal A, Shibli MA (2013) Survey on secure live virtual machine (VM) migration in cloud. In: Conference Proceedings – 2013 2nd National Conference on Information Assurance, NCIA 2013, No. VM, pp 101–106

32. Wang W, Zhang Y, Lin B, Wu X, Miao K (2010) Secured and reliable VM migration in personal cloud. In: ICCET 2010 – 2010 international conference on computer engineering and technology, proceedings, vol 1, pp 705–709

33. Cheng Y (2013) Guardian: hypervisor as security foothold for personal computers

34. Perez R, Sailer R, van Doorn L (2005) vTPM: virtualizing the trusted platform module. UsenixOrg:305–320

35. Mukhtarov M, Miloslavskaya N, Tolstoy A (2011) Network security threats and cloud infrastructure services monitoring. In: ICNS 2011, the seventh international conference on networking and services, No. C, pp 141–145

36. Zhang F, Chen H (2013) Security-preserving live migration of virtualmachines in the cloud. J Netw Syst Manag 21(4):562–587

37. Melliar-Smith PM, Moser LE (2010) O-ring: a fault tolerance and load balancing architecture for peer-to-peer systems. In: Proceedings – international conference of the Chilean Computer Science Society, SCCC, pp 25–33

38. Pantazoglou M, Tzortzakis G, Delis A (2016) Decentralized and energy- efficient workload management in enterprise clouds. IEEE Trans Cloud Comput 4(2):196–209

39. Loreti D, Ciampolini A (2014) A decentralized approach for virtual infrastructure management in cloud datacenters. Int J Adv Intell Syst, Citeseer 7(3):507–518

40. Feller E, Morin C, Esnault A (2012) A case for fully decentralized dynamic VM consolidation in clouds. In: CloudCom 2012 – proceedings: 2012 4th IEEE international conference on cloud computing technology and science, pp 26–33

41. Wang XY, Liu XJ, Fan LH, Jia XH (2013) A decentralized virtual machine migration approach of data Centers for cloud computing. Math Probl Eng 2013:10

42. Sabahi F (2012) Secure virtualization for cloud environment using hypervisor-based technology. Int J Mach Learn Comput 2(1):39

43. Pandya S, Patel W, Ghayvat H (2018) NXTGeUH: Ubiquitous Healthcare System for vital signs monitoring & falls detection. In: IEEE international conference, December. Symbiosis International University

44. Ghayvat H, Pandya S (2018) Wellness sensor network for modeling activity of daily livings – proposal and off-line preliminary analysis. In: IEEE international conference, December. Galgotias University, New Delhi

45. References

46. Xianqin LX, Xiaopeng G, Han W, Sumei W (2011) Application-transparent live migration for virtual machine on network security enhanced hypervisor. China Commun 8(3):32–42

47. Benali R, Teyeb H, Balma A, Tata S, Ben Hadj-Alouane N (2016) Evaluation of traffic-aware VM placement policies in distributed cloud using Cloud Sim. In: Proceedings – 25th IEEE international conference on enabling technologies: infrastructure for collaborative enterprises, WETICE 2016, pp 95–100

48. Bagheri Z, Zamanifar K (2014) Enhancing energy efficiency in resource allocation for real-time cloud services. In: 2014 7th International symposium on telecommunications, IST 2014, pp 701–706

49. Ferdaus MH, Murshed M, Calheiros RN, Buyya R (2017) An algorithm for network and data-aware placement of multi-tier applications in cloud data centers. J Netw Comput Appl 98(September):65–83

50. Nikzad S (2016) An approach for energy efficient dynamic virtual machine consolidation in cloud environment. Int J Adv Comput Sci Appl 7(9):1–9
51. Wen W-T, Wang C-D, Wu D-S, Xie Y-Y (2015) An ACO-based scheduling strategy on load balancing in cloud computing environment. In: 2015 Ninth international conference on frontier of computer science and technology, pp 364–369
52. Zhao Y, Huang W (2009) Adaptive distributed load balancing algorithm based on live migration of virtual machines in cloud. In: 2009 Fifth international joint conference on INC, IMS and IDC, pp 170–175
53. Fu X, Zhou C (2015) Virtual machine selection and placement for dynamic consolidation in cloud computing environment. Front Comp Sci 9(2):322–330
54. Pandya S, Patel W, Mistry V (2016) i-MsRTRM: developing an IoT based iNTELLIGENT Medicare system for real-time remote health monitoring. In: 8th IEEE international conference on computational intelligence and communications networks (CICN-2016), 23–25th December, Tehari, India
55. Pandya S, Patel W (2016) An adaptive approach towards designing a smart health-care real-time monitoring system based on IoT and data mining. In: 3rd IEEE international conference on sensing technology and machine intelligence (ICST-2016), November, Dubai
56. Pandya S, Dandvate H (2016) New Approach for frequent item set generation based on Mirabit Hashing Algorithm. In: IEEE international conference on inventive computation technologies (ICICT), 26 August, India
57. Arianyan E, Taheri H, Sharifian S (2016) Multi target dynamic VM consolidation in cloud data centers using genetic algorithm. J Inf Sci Eng 32(6):1575–1593
58. Mohammed B, Kiran M, Awan IU, Maiyama KM (2016) Optimising fault tolerance in real-time cloud computing IaaS environment. In: Proceedings – 2016 IEEE 4th international conference on future Internet of Things and cloud, FiCloud 2016, pp 363–370
59. Al-Jaroodi J, Mohamed N, Al Nuaimi K (2012) An efficient fault-tolerant algorithm for distributed cloud services. In: Proceedings – IEEE 2nd symposium on network cloud computing and applications, NCCA 2012, No. 2, pp 1–8
60. Jhawar R, Piuri V (2013) Fault tolerance and resilience in cloud computing environments. Comput Inf Secur Handbook 2:125–141
61. Shwethashree A, Swathi DV, Prof A (2017) A brief review of approaches for fault tolerance in distributed systems. Int Res J Eng Technol (IRJET) 4(2):77–80
62. Zheng Q (2010) Improving MapReduce fault tolerance in the cloud. In: Proceedings of the 2010 IEEE international symposium on parallel and distributed processing, workshops and Phd forum, IPDPSW 2010, vol I
63. Kochhar D, Kumar A, Hilda J (2017) An approach for fault tolerance in cloud computing using machine learning technique. Int J Pure Appl Math 117(22):345–351
64. Shribman A, Hudzia B (2013) Pre-copy and post-copy VM live migration for memory intensive applications. In: Lecture notes in computer science (including subseries Lecture notes in artificial intelligence and lecture notes in bioinformatics), vol 7640, LNCS, pp 539–547
65. Rathod SB, Reddy VK (2017) Ndynamic framework for secure VM migration over cloud computing. J Inf Process Syst 13(3):476–490
66. Pandya S, Vyas D, Bhatt D (2015) A survey on various machine learning techniques. In: International conference on emerging trends in scientific research (ICETSR-2015), ISBN No: 978-81-92346-0-5
67. Pandya S, Wandra K, Shah J (2015) A hybrid based recommendation system to overcome the problem of sparcity. In: International conference on emerging trends in scientific research, December
68. Deshpande U, Keahey K (2015) Traffic-sensitive live migration of virtual machines. In: 2015 15th IEEE/ACM international symposium on cluster, cloud and grid computing, pp 51–60
69. Amin Z, Singh H, Sethi N (2015) Review on fault tolerance techniques in cloud computing. Int J Comput Appl 116(18):11–17
70. https://www.ibm.com/blogs/cloud-computing/2014/01/cloud-computing-defined-characteristics-service-levels/

71. Rathod SB, Krishna Reddy V (2018) Decentralized predictive secure VS placement in cloud environment. J Comput Sci 14(4):396–407
72. Teng F, Yu L, Li T, Deng D, Magouls F (2017) Energy efficiency of VM consolidation in IaaS clouds. J Supercomput 73(2):782–809
73. https://github.com/Azure/AzurePublicDataset/blob/master/AzurePublicDataset
74. Jayamala R, Valarmathi A (2021) An enhanced decentralized virtual machine migration approach for energy-aware cloud data centers. Intell Autom Soft Comput 27(2):347–358
75. Patil S, Pandya S (2021) Forecasting dengue hotspots associated with variation in meteorological parameters using regression and time series models. Front Public Health 9:798034. https://doi.org/10.3389/fpubh.2021.798034
76. Sangar D, Upreti R, Haugerud H, Begnum K, Yazidi A (2020) Stable marriage matching for homogenizing load distribution in cloud data center in transactions on large-scale data- and knowledge-centered systems XLV. In: Lecture notes in computer science, vol 12390. Springer, Berlin, Heidelberg
77. Swarna Priya RM, Bhattacharya S, Maddikunta PKR, Somayaji SRK, Lakshmanna K, Kaluri R et al (2020) Load balancing of energy cloud using wind driven and firefly algorithms in internet of everything. J Parallel Distrib Comput 142:16–26
78. Wang T, Quan Y, Shen XS, Gadekallu TR, Wang W, Dev K (2021) A privacy-enhanced retrieval technology for the cloud-assisted Internet of Things. IEEE Trans Industr Inform 18(7):4981–4989
79. Rupa C, Srivastava G, Gadekallu TR, Maddikunta PKR, Bhattacharya S (2020) A blockchain based cloud integrated IoT architecture using a hybrid design. In: International conference on collaborative computing: networking, applications and worksharing, October. Springer, Cham, pp 550–559
80. Vashishtha M, Chouksey P, Rajput DS, Reddy SR, Reddy MPK, Reddy GT, Patel H (2021) Security and detection mechanism in IoT-based cloud computing using hybrid approach. Int J Internet Technol Secur Trans 11(5–6):436–451
81. Prabadevi B, Deepa N, Pham QV, Nguyen DC, Reddy T, Pathirana PN, Dobre O (2021) Toward blockchain for edge-of-things: a new paradigm, opportunities, and future directions. IEEE Internet Things Mag 4(2):102–108
82. Gadekallu TR, Pham QV, Nguyen DC, Maddikunta PKR, Deepa N, Prabadevi B et al (2021) Blockchain for edge of things: applications, opportunities, and challenges. IEEE Internet Things J 9(2):964–988
83. Wang W, Qiu C, Yin Z, Srivastava G, Gadekallu TR, Alsolami F, Su C (2021) Blockchain and PUF-based lightweight authentication protocol for wireless medical sensor networks. IEEE Internet Things J 9(11):8883–8889
84. Krishnan SSR, Manoj MK, Gadekallu TR, Kumar N, Maddikunta PKR, Bhattacharya S et al (2020) A blockchain-based credibility scoring framework for electronic medical records. In: 2020 IEEE Globecom workshops (GC Wkshps), December. IEEE, pp 1–6
85. Somayaji SRK, Alazab M, Manoj MK, Bucchiarone A, Chowdhary CL, Gadekallu TR (2020) A framework for prediction and storage of battery life in IoT devices using DNN and blockchain. In: 2020 IEEE Globecom workshops (GC Wkshps), December. IEEE, pp 1–6
86. Mubashar A, Asghar K, Javed AR, Rizwan M, Srivastava G, Gadekallu TR et al (2021) Storage and proximity management for centralized personal health records using an ipfs-based optimization algorithm. J Circuits Syst Comput:2250010
87. Maddikunta PKR, Pham QV, Prabadevi B, Deepa N, Dev K, Gadekallu TR et al (2021) Industry 5.0: a survey on enabling technologies and potential applications. J Ind Inf Integr:100257
88. Hakak S, Khan WZ, Gilkar GA, Assiri B, Alazab M, Bhattacharya S, Reddy GT (2021) Recent advances in blockchain technology: a survey on applications and challenges. Int J Ad Hoc Ubiquitous Comput 38(1–3):82–100
89. Manoj M, Srivastava G, Somayaji SRK, Gadekallu TR, Maddikunta PKR, Bhattacharya S (2020) An incentive based approach for COVID-19 planning using blockchain technology. In: 2020 IEEE Globecom workshops (GC Wkshps), December. IEEE, pp 1–6

90. Kumar P, Kumar R, Srivastava G, Gupta GP, Tripathi R, Gadekallu TR, Xiong N (2021) PPSF: a privacy-preserving and secure framework using blockchain-based machine-learning for IoT-driven smart cities. IEEE Trans Netw Sci Eng 8(3):2326–2341
91. Kumar R, Kumar P, Tripathi R, Gupta GP, Gadekallu TR, Srivastava G (2021) Sp2f: a secured privacy-preserving framework for smart agricultural unmanned aerial vehicles. Comput Netw 187:107819
92. Ch R, Srivastava G, Gadekallu TR, Maddikunta PKR, Bhattacharya S (2020) Security and privacy of UAV data using blockchain technology. J Inf Secur Appl 55:102670
93. Deepa N, Pham QV, Nguyen DC, Bhattacharya S, Prabadevi B, Gadekallu TR et al (2020) A survey on blockchain for big data: approaches, opportunities, and future directions. arXiv preprint arXiv:2009.00858

Correction to: Federated Learning Enabled Edge Computing Security for Internet of Medical Things: Concepts, Challenges and Open Issues

Gautam Srivastava, Dasaradharami Reddy K., Supriya Y., Gokul Yenduri, Pawan Hegde, Thippa Reddy Gadekallu, Praveen Kumar Reddy Maddikunta, and Sweta Bhattacharya

Correction to:
Chapter 3 in: G. Srivastava et al. (eds.), *Security and Risk Analysis for Intelligent Edge Computing*, Advances in Information Security 103,
https://doi.org/10.1007/978-3-031-28150-1_3

In Chapter 3 "Federated Learning Enabled Edge Computing Security for Internet of Medical Things: Concepts, Challenges and Open Issues," the author name was inadvertently published as "Dasaratha Rami Reddy K". The name has now been updated as "Dasaradharami Reddy K" in the book.

The updated original version for this chapter can be found at
https://doi.org/10.1007/978-3-031-28150-1_3

Correction to: Blockchain Based Simulated Virtual Machine Placement Hybrid Approach for Decentralized Cloud and Edge Computing Environments

Suresh Rathod, Rahul Joshi, Sudhanshu Gonge, Sharnil Pandya, Thippa Reddy Gadekallu, and Abdul Rehman Javed

Correction to:
Chapter 12 in: G. Srivastava et al. (eds.), *Security and Risk Analysis for Intelligent Edge Computing*, **Advances in Information Security 103,**
https://doi.org/10.1007/978-3-031-28150-1_12

The original version of Chapter 12 was published with an incorrect affiliation

Symbiosis Institute of Technology, Symbiosis International (Deemed) University, Mulshi, Maharashtra, India

The affiliation has now been corrected as

Symbiosis Institute of Technology, Pune, Symbiosis International University (Deemed University), Pune, India

The updated original version for this chapter can be found at
https://doi.org/10.1007/978-3-031-28150-1_12

Printed in the United States
by Baker & Taylor Publisher Services